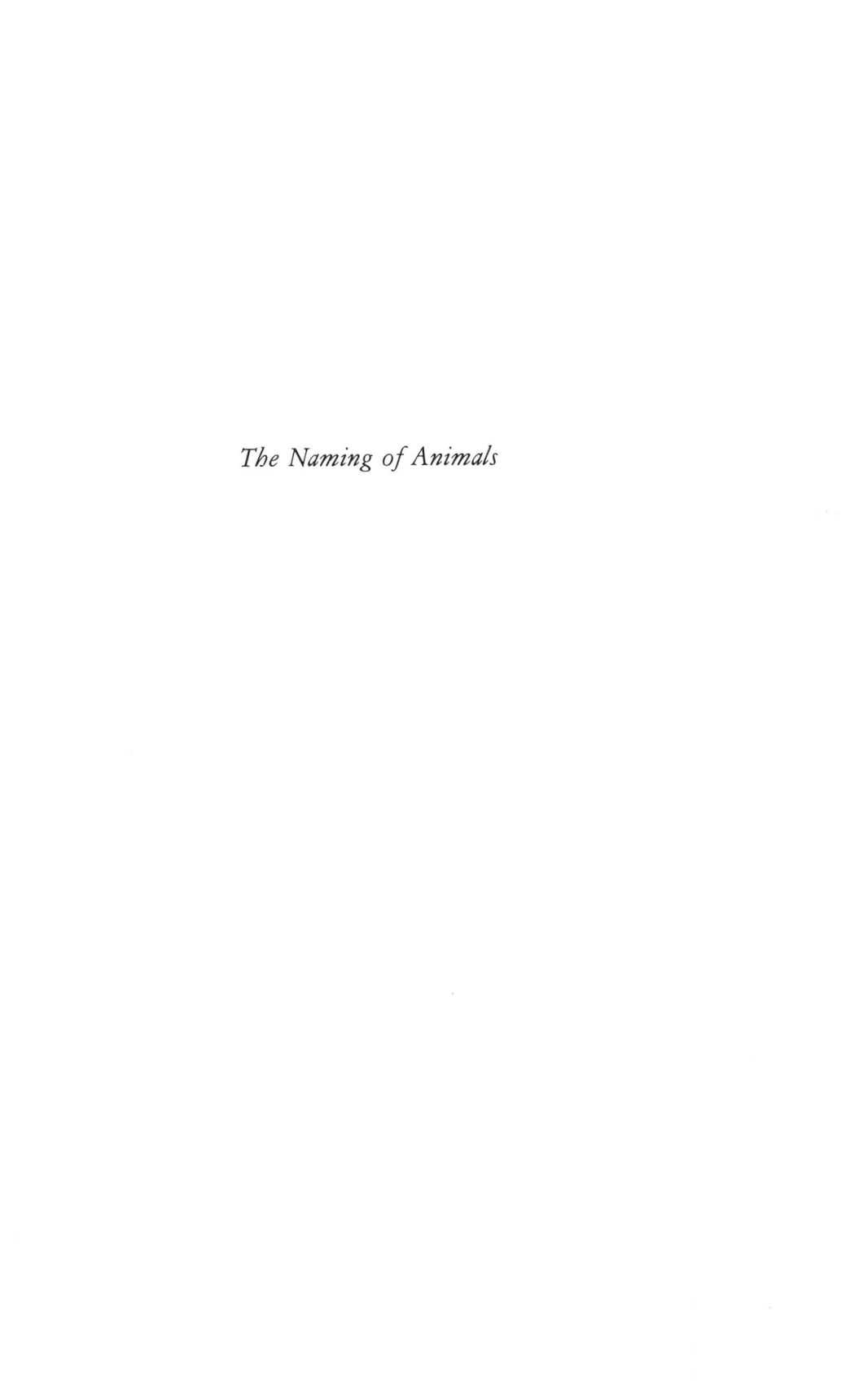

*The Naming of Animals*

# THE NAMING OF
# ANIMALS

AN APPELLATIVE REFERENCE
TO DOMESTIC, WORK
AND SHOW ANIMALS
REAL AND FICTIONAL

*by* Adrian Room

McFarland & Company, Inc., Publishers
*Jefferson, North Carolina, and London*

Acknowledgment is made to the estate of the author and to
Popular Dogs as publisher for permission to reproduce the list of
Beagle names from Thelma Gray, *The Beagle,* Popular Dogs, 1980.

British Library Cataloguing-in-Publication data are available

Library of Congress Cataloguing-in-Publication Data

Room, Adrian.
    The naming of animals : an appellative reference to domestic, work
and show animals real and fictional / by Adrian Room.
      p.   cm.
    Includes bibliographical references (p. 211) and index. ∞
    ISBN 0-89950-795-6 (lib. bdg. : 50# alk. paper)
    1. Domestic animals — Names.  2. Animals — Nomenclature (Popular)
I. Title.
SF21.5.R66   1993
636 — dc20                                                        92-56689
                                                                                      CIP

©1993 Adrian Room. All rights reserved

Manufactured in the United States of America

*McFarland & Company, Inc., Publishers*
  *Box 611, Jefferson, North Carolina 28640*

This book is for Jo,
who as a teenage schoolgirl
responded to a request of mine
for names of horses and
who has been a friend ever since

# CONTENTS

*Acknowledgments*     ix
*Introduction*     xi

| | | |
|---|---|---|
| 1 | Animals and Their Names | 1 |
| 2 | Generic Names | 13 |
| 3 | Descriptive Names | 23 |
| 4 | Incident Names | 37 |
| 5 | Link Names | 45 |
| 6 | Group Names | 57 |
| 7 | Pedigree and Show Names | 67 |
| 8 | Horse and Hound Names | 81 |
| 9 | Farm Animals | 97 |
| 10 | Famous Real Animals | 105 |
| 11 | Animals of Myth, Legend, and Cartoon | 133 |
| 12 | Animals in Fiction | 141 |

*Appendices*

| | | |
|---|---|---|
| I. | Breed Names | 167 |
| II. | Hound Names | 185 |
| III. | Literary Listings | 190 |
| | 1. The Hounds of Actaeon | 190 |
| | 2. Animals in Scott | 191 |
| | 3. Hounds in Surtees | 192 |
| | 4. Horses in Surtees | 193 |

|     |                                      |     |
| --- | ------------------------------------ | --- |
|     | 5. Animals in Dickens                | 194 |
|     | 6. Animals in Dostoyevsky            | 195 |
|     | 7. Animals in Beatrix Potter         | 196 |
|     | 8. Animals in "Saki"                 | 197 |
|     | 9. Animals in Wodehouse              | 199 |
| IV. | Zoo Names                            | 200 |
| V.  | Celebrity Pets                       | 204 |
| VI. | Movie Titles Including Animal Names  | 207 |

*Bibliography*   211

*Index*   215

# ACKNOWLEDGMENTS

This book has been a fairly long time in the making, and over the years I have been indebted to a number of people who have told or sent stories and information about animal names in response to my request, often as the result of a letter of mine that appeared in such magazines as *Cats, Horse and Pony,* and *Our Dogs.* It is well nigh impossible to list every informant individually, but there are some whom I simply cannot omit to thank here by name.

Marian Rickerby was one of the earliest and most prolific providers of horse names, including those of model horses owned by members of the Model Horse Society, which she ran for many years. No model horse names appear in this book, but the reader may be assured that they are very close in nature to those of their real-life bigger brethren. More recently, I owe thanks to five people for information on the names of "professional" horses: to Sally Downes, assistant public relations manager of the Jockey Club, London, to A.J. Crutchley and P.H. Jones, respectively racing manager and stud book manager at Weatherbys, the Jockey Club agents who publish the weekly *Racing Calendar* and biennial *Registered Names of Horses,* to Chief Superintendent G. Fleming, of the Metropolitan Police Mounted Branch, for the names of London police horses and for information on those names, and to J.H. Lawless, foreman horsekeeper at Whitbread's Brewery, London, for the names of the famous "Whitbread Shires."

For dog names, I am likewise indebted to five people. First, to Leslie Dunkling, who allowed me to draw on material that he had begun to compile for a book of dog names that he had planned but that in the event never materialized. Second, to Teresa Slowik, librarian at the Kennel Club, London, who granted me full access to all the books and publications in the Library, and who was more than helpful in providing copies of lists and in answering queries. Third, to Sue Wiseman of the Dickens House Museum, London, who guided me in the right direction for information on Dickens' dogs, both real and fictional, and who also answered queries. For suggestions regarding the origins of some of the names of Sir Walter Scott's dogs, I am indebted to James C. Corson, honorary librarian at Abbotsford House, Scott's famous last home near Melrose, Scotland. Finally I wish to thank Chief Inspector I.R. Hunt, of the Metropolitan Police Dog Training Establishment, Kent, for a comprehensive listing of police dog names, a distillation of which appears in Chapter 8.

For cat names, my thanks are due in particular to Lesley Pring, honorary secretary of the Governing Council of the Cat Fancy, who together with members of her staff provided much useful information, especially on the registered names of cats, and to Group Captain H.E. Boothby, director of the Cats Protection League, who advised on cat names in general and who, also with the assistance of his staff, likewise answered many queries. For the results of two comprehensive surveys of cat names I am indebted first to Spillers Top Cat and Pam Nelson of Spillers Foods, and second to Spillers Kattomeat and Anne Dymock of Charles Barker Lyons, Ltd.

All along I have been encouraged by the author and journalist Celia Haddon, herself a writer on animals (see the Bibliography, beginning on page 211), whose article "Cat-calling" in the London *Daily Telegraph* of March 9, 1991, centered on my research for this present book and resulted in a further "wave" of name stories, most of them relating to cats. I have incorporated several and am, as ever, grateful to my correspondents and to Ms. Haddon for publicizing my request for information.

At one stage, when I was living in Salisbury, Wiltshire, I coopted classes of schoolchildren at various local schools to tell me how they came to name their pets and animals. I owe thanks to them and to their teachers for providing many interesting firsthand accounts. The schools concerned were Cathedral School, Salisbury; Godolphin School, Salisbury; Red House School, Salisbury; St. Edmund's Secondary Modern School for Girls, Laverstock; Wilton C.E. Secondary Modern School, Wilton.

But it is to those literally hundreds of individual informants that I am first and foremost grateful, since without their stories there could hardly have been any book like this. I may not give their own names here, but I have at least included the names of many of their animals.

Stamford, Lincolnshire                                          Adrian Room
August 1993

# INTRODUCTION

We have all had something to do with animals, whether privately or professionally. Privately, we can become passionately devoted to our pets. Professionally, we can spend our lives rewardingly working with animals. We can be as devastated when an animal dies as if it had been our nearest and dearest relative. In 1987, the lament of the British journalist Kate Wharton in the *Mail on Sunday* on the death of Charlotte, her beloved Jack Russell terrier, at the advanced age of 17, resulted in the biggest logging of mail ever received for a single article in that newspaper. There can be few of us who have not "loved and lost" an animal friend in this way.

So because animals play such an important role in our lives, we give them names, just as we give names to the children who are born into our families. Well, not in *quite* the same way, since we are often less inhibited, more adventurous, when it comes to the naming of animals. We are not restricted to the standard stock of first names that we normally draw on for naming a baby. We can coopt ordinary words as names, invent names, and be as original or as unoriginal as we please. We can call one animal *Blackie*, and another *Bilbo*, and only we, or those to whom we confide the information, will know where the latter name came from. Indeed, we can call a snow-white animal *Blackie* if we choose. Few people will seriously demur, and the animal itself is hardly likely to object.

In short, the naming of animals defies conventional linguistic parameters. Even so, like any naming system, it does have its traditional codes and unwritten rules and precedents. And when it comes to the naming of "professional" animals, such as pedigree dogs or racehorses, it has its *written* rules, mainly prescriptive or proscriptive.

This book is a detailed study of the sorts of names we give animals, examining the subject methodically: not by species of animal but by type of name, starting with the general and proceeding to the specific.

Hundreds of animal owners or people working with animals were asked how they came to choose the name for their particular animal, and their accounts are recorded here. The owners range from young children, who tell how they chose a name for their kitten or hamster, to retired folk, who reflect on the names of dogs and cats they have known and loved in the past.

The book does not confine its study of animal names to those of real animals, but also embraces the names of fictional animals, from those found in classical mythology to the animals that still regularly appear in modern fiction and in movies.

The term "animal," incidentally, is taken loosely to apply to almost any nonhuman animate creature, from dogs and cats to reptiles, birds, and fishes. Even a few insects get a look in. So does a dinosaur. Zoo animals feature, too, as do wild animals, who may not have been owned or tamed by human hand but who have nevertheless been given a name.

When it comes to "literary" animals, the division is into real and fictional. Real animals, such as MODESTINE in R.L. Stevenson's *Travels with a Donkey,* are the subject of Chapter 10. Fictional animals, such as Rudyard Kipling's tiger SHERE KHAN, in *The Jungle Book,* find their place in Chapter 12.

The author happens to be British, and has therefore gathered much of his material on the eastern side of the Atlantic. But the traditions and principles of animal naming are much the same throughout the English-speaking world. The names considered herein are chiefly British and American.

There have been many books on animals, but only a few on animal names, mostly by way of recommendatory lists. This book is unique in that it not only considers the *actual* names of animals, many of them familiar, some less so, but has the express aim of giving their histories and origins. The names themselves were taken either directly from a wide range of informants, as mentioned, or from printed sources, the most important of which are listed in the Bibliography beginning on page 211.

His name was Shadow, short for Shadow That Comes in Sight, an old Indian name, Apache or Cheyenne. I very much approved of this. You don't want dogs called Spot or Pooch. You don't want dogs called Nigel or Keith. The names of dogs should salute the mystical drama of the animal life. *Shadow* — that's a *good* name.
(Martin Amis, *Money,* 1984.)

# 1

# *Animals and Their Names*

We humans have long seen animals in anthropomorphic terms, as different but nevertheless clearly defined shadows of ourselves, whether they are farm animals, zoo animals, or the most intimate and dearly loved family pet. Generically we talk of animals as "lesser brethren" or as "four-footed friends," and specifically we have come to think of a dog, that intelligent and faithful animal, as "man's best friend." We individualize animals, too, and just as we name our own children, so we name our animals. In calling our dog SANDY or our cat TIMMY we are following, albeit at a lowlier level, in the footsteps of Adam, who in the biblical story "gave names to all cattle, and to the fowl of the air, and to every beast of the field."

This book is about the names that we give the animals in the historical world, from the chargers of Roman times to the prizewinners of the most recent dog or cat show. The book considers not only the names of real animals, but those of mythology and fiction, many of which have become just as famous as their living equivalents. Indeed, fact and fiction have blended so closely in some cases, that we may not be able to say immediately whether BUCEPHALUS was the name of a real horse or a mythical one, whether LUATH actually existed as a dog or remains a character of Irish legend. In fact, in the latter case, he was both, since the mythical name was adopted for more than one flesh and blood dog. Not only that, but the names of real animals, whether famous in their own right or as private pets, have frequently been adopted for fictional use.

In actual fact the names of fictional and real-life animals are surprisingly similar, and share identical features and characteristics, just as their human imaginary or real counterparts do. In that sense, therefore, the task of this book is not so much to differentiate between the fictional animals and those that actually existed, but to consider their names by comparison with our own.

It will be seen that many animals are given standard personal names, from LENNIE and LISA to PETER and POPPY. There is a marked tendency, however, to give animals diminutive names, "pet names for pets," as it were. This is because animals, even those in zoos and on farms, are frequently regarded as "children." Many animals, too, are physically smaller than we humans, so seem to require an appropriately "smaller" name. It may be significant that of the eight names so far quoted in this chapter, two are genuine diminutives and no less than four have the -IE or -Y ending that often marks a diminutive name. That is, SANDY and POPPY *sound* like diminutives, even though etymologically they are not.

But animal names differ chiefly from human names in their wide range and diversity of origins. It is perfectly acceptable to adopt a standard word and, whether in a diminutive form or not, bestow it on an animal. Many such names are descriptive, or what are explained in Chapter 4 as "incident names," which are really themselves a type of descriptive name. This means that an animal associated in some way with an inanimate object (a true association of opposites!) may well be named for that object. A pup born in a box may be called BOXY, and a kitten that knocks over a cup may get to be called CUPPY.

These two names are what might be called "nonstandard inanimate diminutives," since in English we do not normally use such forms for inanimate objects (we usually say "little" instead). The high representation of diminutives of any kind is also due to the fact that many pets are those of children, and are named by children, whose own world of names involves many diminutives, both for themselves and for their siblings and friends.

A special attribute of animal names is that with the exception of "professional" animals such as racehorses, show dogs and cats, and zoo animals, they are in many cases rarely written. This means that where a human male or female may settle for a particular spelling and stick to it, such as *Jenni* rather than *Jenny*, *Jimi* instead of *Jimmy*, such niceties need not concern the animal namer. Or animal *caller*, of course, since an animal's name performs the important function of summoning it, distracting it, reprimanding it, or at any rate attracting its attention. This further factor may, at least subconsciously, affect the sort of name we give an animal. We want one that is going to be easy to say (or call) and that will be readily heard and responded to.

Curiously, few of the many books on the raising, breeding, and general care of animals offer advice about naming, even in its practical aspects. However, the subject is considered by the Canadian-born veterinarian Bruce Fogle in *Pets and Their People* (see Bibliography), where he reinforces the point that male animals are usually given "strong and active" names, but females are given "passive and friendly" names, both types thus conforming to the sexist norm traditionally associated with humans. At the same time, common sense dictates that an animal's name should be suitable for its breed and species, and that, in the case of animals that are required to respond to commands, the name should not be one that could be confused with such a command. It could be disconcerting for a dog named *Sid* to be summoned by name one moment, then given the command "Sit!" the next. The same could apply when a dog called *Stacy* is given the order "Stay!"

It is always interesting to see how the experts choose their names. The British dog trainer and television personality Barbara Woodhouse, for example, has described in her book *Almost Human* (1981) how she had a Great Dane named JUNO and how she came to name her successor, JUNIA: "I couldn't use the name JUNO again, it was sacred to that lovely dog now gone. I had to choose a name that meant something and this surely did. I always shortened it to Juni because the 'i' at the end made it easy for the puppy to identify herself with that name. Junia would have had one syllable too many for quick communication."

It might be thought that many people would aim to choose an entirely new or original name for an animal. It is clear, however, that this is not the case. Indeed, not only do owners often give traditional names, whether personal in origin or not, but in many instances give a name that is random or arbitrary. It is understandable, of course, that farmers, who deal with hundreds of animals in their lives, should give stock names, that such names should be repeated, and that the namers should at times be hard pressed to think of a new name at all for a particular animal (see Chapter 3). The difficulty is alluded to by the British naturalist and novelist W.H. Hudson, in what is generally regarded as his finest book, *A Shepherd's Life* (1910), where he comments on the naming of sheepdogs:

> On receiving the pup he [the shepherd Caleb Bawcombe] was told that its name was Tory, and he did not change it. It was always difficult, he explained, to find a name for a dog — a name, that is to say, which anyone would say was a proper name for a dog and not a foolish name. One could think of a good many proper names — Jack and Watch, and so on — but in each case one would remember some dog which had been called by that name, and it seemed to belong to that particular well-remembered dog and to no other, and so in the end because of this difficulty he allowed the name to remain.

The problem can also exist in domestic circles, where an animal is simply there as a pet, and where his or her name is of minor or even no importance, even though there may be a regular requirement to call the animal or at least attract its attention. The thinking here may be on the lines of: "Animals are simple creatures, their names mean nothing to them, so let's give any old name." But do the names mean nothing to the namers? Parents who name their child *Louise* or *Martin* often do so because the name is meaningful to *them* in some way, and because they have now given it as a gift to their offspring, who will constantly bear it and in a real sense embody it. The name may lack any special significance to the child, but he or she will nevertheless be constantly aware of it and promote it. The same is true of animals. And when we meet a strange dog, one of the first questions we often ask his owner is, "What's his name?"

Some names, on the other hand, are positively planned, however speedily, with a due process of consideration, rejection, and adoption. The television personality and popular writer Daniel Farson tells in his book *In Praise of Dogs* (see Bibliography) how he came to call his dog LITTLEWOOD. Some boys had originally bought the dog in a London street market, and subsequently came to Farson with her:

> "What's she called?" I asked weakly. There was a sigh of relief and their eyes became confident.
> "Her name's Trix. Trixie, ain't it?"
> "Oh no it isn't," I explained, thinking desperately of a suitable feminine name. I had been working all the day with Barbara Windsor and Joan Littlewood on a film called *Sparrows Can't Sing*. 'Barbara' sounded silly for a dog, so did 'Joan' for that matter.
> "She's going to be called Littlewood," I declared.
> "That's not a dog's name," said one boy scornfully, "why call her that?"
> "Because of the pools," explained Rose, mistakenly but prophetically.
> And that is how Littlewood entered my life and came to be part of it.

Presumably this name was abbreviated for use as a pet or call name. (Littlewoods is a leading British football betting company; Rose was Farson's maid.)

The naming of a fictional animal can take place on similar lines. Here, a century earlier, is the naming of the stray dog WAIF, as described in Edna Lyall's novel *Donovan* (1882):

> "What shall we call him? Harlequin?"
> "No, that's too long, and it must mean something that's lost and all alone," said Dot. "Rover would do, only it's so common."
> "Vagabond, Tramp, Waif or Stray," suggested Donovan.
> "Oh—Waif—that's beautiful, and so nice to say."
> "Yes, a thing tossed up by chance; it'll just suit the beggar."

## 1. Animals and Their Names

A novelist who is also an animal owner may need to consider carefully what names to use when including an animal in a fictional work. Should the animal retain its own name? If not, what should its new name be? The Irish writer Forrest Reid has described his own method of solving the problem, with regard to his bulldogs PAN and REMUS, his Irish terrier NYX, and his sheepdog ROGER:

> I never wrote about Nyx, I never wrote about Roger, but I wrote about Pan in *Following Darkness*, and Remus in *The Spring Song*. I gave them other names. I called Pan Tony, and Remus Pouncer. Pouncer was a good name, but Tony wasn't, so when I rewrote the book as *Peter Waring* I changed it. With the wrong name I can make no headway. (*Private Road*, 1940.)

Reid in fact changed TONY back to PAN. His observation that he could "make no headway" with the wrong name is significant here. He was as selective in his animal names as many writers are over the names of fictional characters.

Many writers who are animal owners or lovers prefer instead to write about the animals under their true names. The books by the British publisher Michael Joseph on his beloved cats (see under his name in Bibliography) are examples of this. Another such book, just one of many, is *All the Dogs of My Life* (1930) by "Elizabeth," otherwise Elizabeth von Arnim, already familiar to her readers as the author of *Elizabeth and Her German Garden* (1898). In this autobiographical account, she tells how she has owned 14 dogs; in order, BIJOU, BILDAD, CORNELIA, INGRABAN, INGULF, INGO, IVO, PRINCE, COCO, PINCHER, KNOBBIE, CHUNKIE, WOOSIE and WINKIE. The fourth to seventh of these were related, with every dog bearing a name beginning with "I." BILDAD was a Pomeranian (as the author herself would later become). The writer was given it at the age of 14 and named it herself: "It was I who christened him Bildad, being at that time an earnest student of the Bible; and when my Aunts Charl and Jessie [...] asked, 'Why Bildad?' I said it was because he was the height of a shoe; and, on their inquiring further, referred them glibly to the Book of Job, second chapter, eleventh verse." The verse quoted names Job's three friends, one of whom is "Bildad the Shuhite." Most of the author's other dogs are also gifts, but came already named.

What are the most common names for dogs? We can do no better than cite the top 50 dog names extracted from the 13,498 dog licenses issued by the Department of Health in Detroit, Michigan, in 1983. They are as follows (with the number of occurrences in brackets):

LADY (274), DUKE (151), PEPPER (143), KING (137), GINGER (128), SHEBA (126), PRINCESS (122), BRANDY (117), PRINCE (114), HEIDI (112), SANDY (112), SAM (111), BUFFY (100), BUTCH (100), MAX (98), RUSTY (98), BLACKIE (92), ROCKY (89), SNOOPY (82), PENNY (81), SMOKEY (79), REX (74), QUEENIE (72), TOBY (71), BENJI (70), CHARLIE (69), SPARKY (69), CANDY (68), CHICO (68), LUCKY (65), TIGER (64), MUFFIN (63), FLUFFY (61), TIPPY (61), BEAR (60), DUCHESS (60), BUTTONS (58), CHAMP (57), MANDY (56), BUDDY (55), MISSY (54), PIERRE (53), TAFFY (53), MICKEY (52), COCO (51), TRIXIE (51), BROWNIE (50), BANDIT (47), BARNEY (47), DUSTY (47).

While this is a good representative sampling, with many of the names doubtless widely found elsewhere in the United States and indeed throughout the English-speaking world, it should be remembered that this is specifically the sampling for Detroit, with its essentially urban and cosmopolitan population, roughly half black, where dogs may be kept as much to protect family and home as to be pets. The fact that LADY is top of the list does not imply that there are more female dogs than male, but that dog namers often like to give a bitch a name that indicates her sex. Hence the other high incidence of female names, such as SHEBA, PRINCESS and HEIDI.

On the other hand, specifically male names also have high ratings, notably the "royals": DUKE, KING and PRINCE. Personal names, such as SHEBA, SAM and MAX, are easily outnumbered by purely descriptive names, especially those denoting colors (PEPPER, GINGER, BRANDY, RUSTY, BLACKIE, SMOKEY, and the rest). Information is lacking with regard to the particular breed that goes with a name. However, DUKE and KING are much more likely to be German shepherds than poodles, while COCO and TRIXIE are probably small or toy dogs.

Across the Atlantic, it would appear that British dog owners have a more marked preference for personal names for their animals, to judge by the top dog names published annually by the London *Daily Telegraph*. The names are provided by a veterinarian from an admittedly much smaller sampling of 1,000 names of dogs treated. For the years stated the most popular names were, in order, as follows (D = male dog, B = bitch):

1983
D BEN, SAM, MAX, WILLIAM, TOBY, CHARLIE, OLIVER, SNOOPY, JASON;
B SOPHIE, EMMA, LUCY, TESSA, GEMMA, ZOE, SUSIE, CANDY, BONNIE, CINDY.

1984
D  BEN, SAM, WILLIAM, JASON, DOUGAL, CHARLIE, SNOOPY, MAX, JAMIE, TOBY;
B  EMMA, GEMMA, TESSA, KATE, ZOE, SOPHIE, RITA, SUSIE, LUCY, KIM.

1985
D  BEN, SAM, WILLIAM, CHARLIE, MAX, JAMIE, HARRY, JASON, TOBY, LUKE;
B  EMMA, GEMMA, KATE, SUSIE, LUCY, RITA, KIM, CHLOE, SALLY, TARA.

1987
D  SAM, BEN, WILLIAM, JAMIE, MAX, JASON, ROSS, PRINCE, OSCAR, MARCUS;
B  GEMMA, EMMA, SUSIE, SALLY, KIM, PENNY, SOPHIE, LUCY, DAISY, KATE.

1988
D  SAM, BEN, WILLIAM, ROSS, PRINCE, JAMIE, MAX, HARRY, JASON, BENGY;
B  EMMA, GEMMA, SUSIE, PENNY, SOPHIE, DAISY, KIM, SALLY, MEG, SHEBA.

Apart from SNOOPY, these names actually mirror the true first names that were the most popular in the stated years. Thus the *Daily Telegraph* itself reported that the most common girls' names recorded in its birth announcements during 1985 were: Sarah, Charlotte, Sophie, Elizabeth, Victoria, Hanna, Catherine, Emma, Laura, Lucy. If we allow KATE as a form of *Catherine* and SALLY as a form of *Sarah,* then four of the top 10 girls' names for 1985 were also bitch names. It could well be, however, that this particular veterinarian specializes in the more genteel kind of pet, as distinct from the robust and urban working dog.

The same newspaper's equivalent listing of leading names for cats shows few if any personal names and a predominance of genderless color names. The paper noted that the top cat names for 1987 included: BLACKIE, SOOTY, SNOWY, GINGER, WHISKY, TABBY, TIGER. Where personal names *were* favored, female cats were mostly LUCY and TABITHA, while the top male cat name, as for the dogs, was SAM. In 1989, by way of contrast, the *Telegraph* reported the results of a similar survey by the American magazine *Cat Fancy.* The most popular names for female cats in the United States were as follows: MISTY, PATCHES, MUFFIN, SAMANTHA, FLUFFY, PUNKIN, MISSY, TABITHA, TIGGER. Top names for male cats were: SMOKEY, TIGER, MAX, CHARLIE, ROCKY, SAMMY, MICKEY, TOBY. Here, the choice is wider and more original, as in most American naming practices, with personal names favored far more than in Britain, and with some names unlikely to be given to British cats at all. PUNKIN, in particular, would never feature in any British "Top 10" listing, if only because the British associate *pumpkin,* the source of PUNKIN, with the large European squash *Cucurbita maxima,* a relative rarity (except at Hallowe'en), not the small American squash *Cucurbita pepo,* found regularly in pumpkin pie.

It will be noticed that all these names pay scant heed to the animal's breed. They could apply to almost any breed of dog or cat, or none at all. Yet breeds have their traditional names, too. A veterinarian writing in the summer of 1973 number of the British quarterly *The Countryman* reported that the most popular names for Alsatians (German shepherds) were SHEBA, SASHA, SIMBA, TARA, ELSA and LISA, while for poodles the favorites were, in this order, PEPI, SUZETTE, MITZI, MICHELLE, FIFI and CHERI. Dachshunds frequently came up as KARL, FRITZ, OTTO, MAX and HEIDI, all obviously German names, although SAUSAGE, TREACLE, SHORTY and TOFFEE also regularly recurred.

There are other types of animals besides dogs and cats, of course, and their names are just as important. Many households have children who own guinea pigs or hamsters, for example, and there are gerbils in gardens and ponies in paddocks. In May 1991 the British RSPCA reported the results of a poll of more than 400,000 youngsters in England and Wales for National Pet Week. The findings included the following "Top 10" favorites, in order of choice:

| *Dogs* | *Cats* | *Goldfish* | *Rabbits* | *Budgies* | *Guinea pigs* |
|---|---|---|---|---|---|
| BEN | SOOTY | JAWS | SNOWY | JOEY | SQUEAK |
| SAM | TIGGER | GOLDIE | THUMPER | BILLY | GINGER |
| LADY | TIGER | FRED | FLOPSY | BLUEY | BUBBLES |
| MAX | SMOKEY | TOM | SOOTY | BOBBY | PATCH |
| SHEBA | GINGER | BUBBLES | SMOKEY | SNOWY | SNOWY |
| TOBY | TOM | GEORGE | PETER | PETER | ROSIE |
| SALLY | FLUFFY | FLIPPER | FLUFFY | CHARLIE | SOOTY |
| LUCY | LUCY | BEN | BUGSY | MAGIC | SANDY |
| BONNIE | SAM | JERRY | ROGER | GEORGE | BUBBLE |
| BENJI | LUCKY | SAM | BLACKIE | TWEETY | FLUFFY |

| *Hamsters* | *Gerbils* | *Tortoises* | *Rats* | *Ponies* | *Mice* | *Stick Insects* |
|---|---|---|---|---|---|---|
| HAMMY | SQUEAK | SPEEDY | RATTY | COPPER | MICKEY | STICKY |
| HONEY | JERRY | FRED | ROLAND | BEAUTY | SQUEAK | FRED |
| HARRY | BUBBLE | TOMMY | SPLINTER | BRAMBLE | JENNY | TWIGGY |
| FLUFFY | TOM | TOBY | BEN | STAR | MINNIE | TOM |
| BUBBLES | SNOWY | TERRY | SQUEAK | MISTY | SPEEDY | GEORGE |
| SNOWY | BUBBLES | TIMMY | BLACKIE | LADY | BUBBLE | STICK |
| NIBBLES | SOOTY | LEONARDO | RAMBO | AMBER | TOM | SAM |

*1. Animals and Their Names* 9

| Hamsters | Gerbils | Tortoises | Rats | Ponies | Mice | *Stick Insects* |
|---|---|---|---|---|---|---|
| JOEY | SANDY | TOM | TOM | SNOWY | GEORGE | BILLY |
| GIZMO | JOEY | DONATELLO | ROSIE | HOLLY | HARRY | FREDDIE |
| HENRY | SWEEP | GEORGE | SOOTY | BISCUIT | BILL | CHARLIE |

The names are mainly unimaginative, it is true, and many are conventional. Even so, the sampling illustrates interesting features and trends. Dogs, as noted, have mainly personal names, while other animals have mostly descriptive. Colors are common, especially for cats and ponies, with SNOWY a stock name for a white animal and SOOTY for a black-coated one. In the latter case, the name is often paired with SWEEP (see below). Several names are derivative, their source being a popular or topical exemplar, mostly cartoon or puppet characters. JAWS is thus named for the eponymous shark in the movie; ROLAND is named for ROLAND RAT, a popular muppet-like character on British television from 1983; MICKEY and MINNIE are for the world famous cartoon mice; SPEEDY the tortoise has a jocular name. SQUEAK appears for four kinds of small animal, as descriptive of their voice; BUBBLE (or BUBBLES) is its "pair," for the popular dish, *bubble and squeak*. However, for the goldfish, BUBBLES is directly descriptive.

There is a regular stock of common first names, mainly male, which is traditionally drawn on for animal use. It is mostly responsible for the animals named BEN, BILLY, CHARLIE, FRED, GEORGE, JOEY, SAM, and TOM. Female stock names similarly are LUCY and ROSIE. FRED and GEORGE, in particular, are normally reserved for the less individualistic type of animal, such as a goldfish, tortoise, or stick insect. FRED, in fact, has become a virtually standard name for any unknown person, creature, or even inanimate object, although it is now itself rarely given or adopted as a personal name. It is not clear why the name came to be used in this way.

It will be noticed that some animals have names that alliterate with their type. Budgerigars are thus BILLY, BLUEY, and BOBBY; hamsters are HAMMY, HONEY, and HARRY; tortoises are TOMMY, TOBY, TERRY, and TIMMY; rats are RATTY, ROLAND, RAMBO, and ROSIE. TERRY, of course, suggests *terrapin,* and some alliterative names were adopted from cartoon or other animals, as mentioned. Other specific allusions in the listings, whether now consciously made or not, include BUGSY for the Disney cartoon rabbit *Bugs Bunny,* FLIPPER for the dolphin in the movie of the same name, FLOPSY for the *Flopsy Bunnies* of Beatrix Potter, LEONARDO and DONATELLO for two of the four turtles in the cult movie series, *Teenage Mutant Ninja Turtles,* PETER for Beatrix Potter's *Peter Rabbit,* RAMBO for the tough hero of the movie of the same name, ROGER for the rabbit in the movie *Who Framed Roger Rabbit* (1988), SPEEDY for the Mexican mouse in the animated cartoon *Speedy Gonzales,* SPLINTER for the rat guru in *Teenage Mutant Ninja Turtles,* THUMPER for the rabbit in the Walt Disney movie *Bambi,* SOOTY and SWEEP for the teddy bear puppets of British television, TWEETY for the canary TWEETY PIE in the Warner Bros. cartoon (with SYLVESTER the cat). Tortoises named FRED may have had their name suggested by the tortoise of this name who appeared as one of the pets in the children's television program *Blue Peter.* (See also PETRA in Chapter 10, page 126.)

It seems highly probable that some animals were renamed for the express purpose of bearing modishly popular names, and this is almost certainly the case with many of the tortoises LEONARDO and DONATELLO, since this particular animal is notoriously long-lived (in some cases for as many as 150 years), and therefore likely to have borne an earlier, less original name.

The influence of popular characters like these seems perennial. Beatrix Potter's *The Tale of Peter Rabbit* originally appeared (as the author's first book) in 1902. It will be interesting, however, to see whether the names of the turtles and rat in *Teenage Mutant Ninja Turtles* will be equally popular for animal naming use ninety years from now! Possibly the hero of *Who Framed Roger Rabbit* as a solo character is more lastingly memorable.

Commenting on its survey, the RSPCA noted that dogs in urban areas, such as London, tended to have names such as BRUNO, ROCKY, or TYSON, all for famous boxers (one British, two American), perhaps indicating the trend in city districts to keep a dog for protection rather than purely as a pet. Dogs in rural areas, by contrast, were usually given names such as BRACKEN, DAISY, or HOLLY. There appeared also to be a high incidence in southeast England of dogs named CHARLIE and PRINCE, possibly suggesting their owners' interest in the royal family. That there are regional influences at work is supported by the high influence of GAZZA as a name for animals in northeast England, home territory of popular footballer Paul Gascoigne, nicknamed *Gazza,* while BOYCOTT was similarly noted in north central England, doubtless given by fans of the world-class Yorkshire cricketer Geoff *Boycott.*

Wild animals may equally be given names. An advertisement placed in the London *Times* of March 16, 1992, by the Whale and Dolphin Conservation Society, a charitable organization, invited readers to "adopt a whale," the choice of four being HOLLY (with her pup IVY), STRIDER, TOP NOTCH (so called from the "notch," or portion missing from the top of his dorsal fin), and SHARKY (named for the sharklike outline of her dorsal fin).

Even these names do not obviously pertain to the particular type of animal, however, and with due allowance for anatomical differences, could equally be suitable for four dogs, four cats, or four horses. (The reader is invited to repeat the names to a friend and ask what animal they designate. The answer will almost certainly be "Horses"!)

Animal names are thus all things to all people. They are both traditional and innovative, unimaginative and original, highly appropriate and totally inconsequential, soberly meaningful and simply crazy. Proof that names of all such types exist, and have existed for centuries, will be found aplenty in the chapters that follow.

# 2

# *Generic Names*

A special characteristic of animals, and particularly pets, is that although they of course have a standard noun to name them, such as *dog, cat, horse, monkey*, and the like, they in many cases have what might be described as a generic name that is more of a real name in the normal sense of the word, or that has become associated with a personal name. Cats, for instance, are called KITTY, monkeys are JACKO, lions are LEO, parrots are POLLY, budgerigars (and formerly caged canaries) are JOEY, horses are DOBBIN, donkeys are NEDDY if male and JENNY if female, goats are BILLY and NANNY similarly, and so on. In some cases a personal name has actually become the standard word for the animal, so that *tom* is used for a male cat, *robin* is the bird originally known as a *redbreast* (compare French *rouge-gorge* as the standard word for it), *jackdaw* has added the familiar first name to what was originally just *daw*, and similarly *magpie* has prefixed *pie* (still the regular French word for this bird) with a pet form of the name *Margaret. Martin*, too, came to be used for the familiar bird of the swallow family. Even *bird* itself, in children's language, becomes personalized as *dicky bird*.

The adoption of personal names for animals like this is an interesting phenomenon that deserves attention.

Generally speaking, it illustrates our inherent desire to see animals in human terms. The particular personal names that came to be associated with each animal in some cases arose through their use in medieval fables. One of the most popular and influential in this respect was the group of versified French fables known as the *Roman de Renart*, composed by different authors some time in the late 12th century, and itself based on tales already current generally in Europe, including a Latin version of about 1148. The work is a satire on human society, with the central theme the struggle for power between the cunning fox REYNARD and the physically powerful wolf ISENGRYM (who usually wins). Other characters in the tales (in Caxton's English translation from a Flemish version) are King NOBLE the lion, BRUIN the bear, TIBERT the cat, COURTOYS the hound, GRYMBERT the badger, COART the hare, BELLYN the ram, MARTIN the monkey, CHANTICLEER the cock, PARTLET the hen, and TIERCELIN the rook.

The fables became so well known in France that REYNARD's name came to give the standard French word *renard* for a fox, replacing the earlier *goupil*, itself ultimately from Latin *vulpes*. The personal name, today familiar in German as *Reinhart*, derives from the Germanic words *ragin*, "advice," and *hard*, "hardy," "brave." The fox is thus as it were a "strong adviser." Although the other names are today less readily associated with their particular animal (with perhaps the exception of BRUIN), their origins are in most cases equally significant.

ISENGRYM has an appropriately "tough" name, from Germanic *īsen*, "iron" (modern German *Eisen*), and *grīm*, "mask." Both wolf and fox are related members of the dog family *Canidae*, and it is no coincidence that *mask* is a term still current in hunting parlance for a fox's face or head. (See also the name of GRYMBERT below.) NOBLE is a self-explanatory name for the lion as "King of the Animals." BRUIN represents Germanic *brun*, "brown," which also lies behind the modern personal name *Bruno*. The brown bear is the most familiar of its family.

The name of TIBERT the cat has its modern equivalent in *Theobald*, from Germanic *theud*, "people," and *bald*, "bold." Cats today may be associated with timidity, but they have long been traditionally regarded as a "daring race," renowned for their curiosity and enterprise (and survivability). It is this name that appears to have given the modern TIBBLES as a common cat name, while *tibcat* was formerly a term for a female cat, corresponding to the male *tomcat*. TIDDLES, however, the popular modern equivalent of this, derives from the verb *tiddle* meaning "to pet," "to pamper," first recorded by the *Oxford English Dictionary* in 1560. The name later became associated with *tiddly* in the sense "very small" or the related *tiddler*, "very small thing." (For a famous individual cat of the name, see Chapter 10, page 130.) The name TIBERT was adopted by Shakespeare (in the form *Tybalt*) for a character in *Romeo and Juliet* (1594). He is punningly referred to by Mercutio (who kills him in a duel) as "rat-catcher," "Prince of Cats," and "King of Cats."

COURTOYS the hound has a name that means "refined," "accomplished," from Old French *courtois*, which also gave the English surname *Curtis* and the modern identical first name. The name implies a well-educated person Hounds (or dogs in general) are noted for their sagacity. GRYMBERT the badger has a name deriving from Germanic *grīm*, "mask," and *berht*, "famous" (literally, "bright"), so that he was the one with the "familiar face." Compare the name of ISENGRYM above.

The name of the hare, COART (or CUWAERT), has its modern equivalent in English *coward,* referring to a timid creature. The name itself directly represents Old French *coart,* "tail" (from Latin *cauda*), and perhaps alludes to the way in which a frightened animal puts its tail between its legs. This is not the case with the hare, however, and here the reference is probably to its tail which is conspicuously visible as it runs off. *Coward* was an actual word for a hare in English in medieval times.

BELLYN the ram has a name from which the modern French word for the animal, *bélier,* directly derives. It represents Old French *belin,* from a root element found in other European words for "ram," for example Russian *baran.* It almost certainly has no connection with English *bell,* despite the fact that the ram, as leader of the flock, traditionally had a bell round its neck. (Hence *bellwether* as a figurative term for a leader.)

MARTIN is still a familiar personal name, made famous by St. *Martin,* the 4th-century French bishop renowned for his charity. The name itself perhaps derives from *Mars,* genitive *Martis,* the Roman god of war, and may have been adopted for the monkey because of the animal's notorious ferocity. In *Reinke de Vos,* a 15th-century Low German version of the *Roman de Renart,* MONEKE is the name of the son of MARTIN the ape, and it is more than likely that his name actually gave the English word *monkey.* MONEKE itself appears to represent the Flemish diminutive of some personal name. It may have been chosen for the ape because of its association with Low German *monnik,* "monk," from the animal's appearance. (Compare French *moineau,* "sparrow," literally "little monk," referring to the bird's brown head.) *Martin* was for a while a general word for a monkey in English down to about the 16th century, when it was superseded by *monkey* itself. It is now thought that the same personal name also lies behind Russian *martyshka,* strictly meaning "marmoset" but also colloquially "monkey." (English *marmoset* is almost certainly not derived from *Martin,* however, although the word may well have been influenced by its "monkey" sense.)

CHANTICLEER the cock has a much more transparent name, meaning "one who sings clearly," from Old French *chanter,* "to sing," and *cler,* "clear." The name has been regularly taken up in English by poets since the time of Chaucer, for example by Shakespeare and Longfellow. PARTLET, as the name of the hen, is a form of the Old French female personal name *Pertelote.* This is itself of uncertain origin, but it subsequently influenced the word *partlet* as a term for a former kind of ruff worn by women. The actual derivation of this word, however, is probably in Old French *patelette,* "little band." TIERCELIN the rook has a name related to English *tercel* as a word for a male hawk. Modern French *tiercelet* is a term for the male of various kinds of bird. Rather unexpectedly, the word has its own origin in Latin *tertius*, "third," perhaps alluding to the fact that some male birds are one-third the size of the female, or to the belief that the third egg in a nest was smaller than the others and produced a male.

This takes us back to the more common names mentioned at the beginning of this chapter, together with others like them.

KITTY as a name for a cat is in fact not connected with the female personal name, which is a pet form of *Katharine*. It thus actually derives from *kit*, a short form of *kitten*, in turn ultimately from a French form of *cat* itself. (Compare modern French *chaton*, "kitten.") On the other hand MOGGIE, as both a name and colloquial word for a cat, is almost certainly a form of *Maggie* in origin.

*Jack* is familiar as a name prefixed to a generic word to denote a male animal. *Jackass* is a well-known example. In the case of *jackrabbit*, however, the name is a shortening of *jackass rabbit*, referring not specifically to a male but to a species of prairie hare found in western North America that is distinguished by its very long ears, like those of a *jackass* or donkey. *Jack* has also been widely recorded as a folk name for birds other than the *jackdaw* already mentioned. In *A Dictionary of English and Folk-Names of British Birds* (1912), the British ornithologist Harry Kirke Swann noted the following local names from various parts of the British Isles: *Jack Bird* for the fieldfare, *Jack Doucker* for the dabchick, *Jack Hawk* for the kestrel, *Jack Hern* for the heron, *Jack Ickle* for the green woodpecker, *Jackie Foster* for the long-tailed duck, *Jack-in-a-bottle* for the long-tailed tit, *Jack Nicker* or *Jack-a-Nickas* for the goldfinch, *Jack Plover* for the dunlin, *Jack Squealer* for the swift, *Jack Straw* for the whitethroat, and *Jacksaw* for the great tit. *Jack Baker* has also been recorded in the south of England as a name for the (now rare) red-backed shrike. It is of course possible that in some cases the name *Jack* was imitative in origin, and was suggested by a bird's harsh cry (compare *chack* or *chackbird* as a local name for the wheatear). The *jackdaw* may well have got its name in this way. However, this particular origin can hardly have produced the names of all birds so called, especially the smaller ones.

Both JACKO and JOCKO are familiar names for a monkey. JACKO may have actually evolved from JOCKO under the influence of the personal name *Jack*, as for the *jackass* just mentioned. The word *jackanapes* may have had its influence, too, as both a term for a mischievous person and, formerly, as an actual word for an ape or monkey from the 16th century. *Jack* has long been widely used of male humans, as in *Jack Tar* for a sailor and *Jack the Ripper* for the long-unidentified murderer of prostitutes of London in 1888. Hence such phrases as "every man jack of you" or "before you could say Jack Robinson." But this leaves us with JOCKO, which looks like an adoption of the Scottish name *Jock*. According to some sources, however, it is a corruption of *idiok*, an African word for a chimpanzee. And although it is tempting to see it as a form of Italian *gioco*, "game," "sport," especially in view of the association between monkeys and Italian organ grinders, this is very likely a folk etymology.

LEO for a lion is an adoption of the Latin word for the animal, so that the personal name *Leo* derives from this, not the other way round.

Why are parrots called POLLY? As a personal name, *Polly* is a form of *Molly*, itself a pet form of *Mary*. It probably came to be given to the parrot through its similarity to the word *parrot* itself. The bird's own name has been popularly derived from the French personal name *Perrot*, a form of *Pierre*, "Peter," but may actually have a different origin. Another diminutive of this French name, the more familiar *Pierrot*, has itself become a generic nickname for a sparrow. This in turn compares with the former use of PHILIP in English as a nickname for that bird. The actual name *Philip* means literally "lover of horses," from the Greek, but this was not the reason for its use for the sparrow, and it was probably adopted purely from its suggestion of the bird's chirping. (Compare *dicky bird*, below.)

The use of JOEY for a canary, and in more recent times a budgerigar, may arise from the application of the name generally to a small child or young animal, as found still in Australia for a young kangaroo. It may be no coincidence that the budgerigar, or parakeet, is native to Australia, so that when the bird was exported its name went with it.

*Philip* meaning "sparrow" was popularized by the English poet John Skelton's poem *Phyllyp Sparowe* (1529), a lamentation by a young lady for her pet sparrow which had been killed by a cat, itself named GIB:

> Nothynge it auayled
> To call Phyllyp agayne,
> Whom Gyb our cat hath slayne.

GIB is no longer in general use for a cat, although *gibcat* is still sometimes found for a male cat, especially one that has been castrated. It is probably an adoption of the pet form of *Gilbert*, although it is uncertain why this particular name should have been adopted. As a cat name it dates from medieval times, and an inscription of about 1400 has been recorded with the words "Gret : wel : gibbe : oure : cat" (*Proceedings of the Society of Antiquaries,* March 11, 1886).

DOBBIN, as a name for a horse, especially a workhorse or carthorse, is an adoption of the now rarely found personal name that is itself a pet form of *Robin* (see above for the latter's use as the standard name of the bird). The adoption appears arbitrary, and does not seem to date much earlier than the 16th century. It is found in Shakespeare.

NEDDY, from the personal name that is a pet form of *Ned*, itself a short form of *Edward*, has been a generic nickname for a donkey since at least the 18th century, perhaps earlier. The adoption of this particular name may have been influenced by *noddy* as a term for a foolish person, or by *noddle* as a word for the head. Donkeys are noted for their apparent obtuseness and for their prominent head, which they also *nod* or bend downward. JENNY, his female counterpart, has a still popular personal name that is a pet form of *Jean*. It is found for some other animals (compare *jackass*, mentioned above), and especially for birds, in particular the *jenny wren*. In the latter case, however, the personal name has not become the regular word for the bird, as it has for the *robin*.

A male goat is a *billy goat* and a female a *nanny goat*. The two names represent personal names that are respectively the pet forms of *Bill*, a short form of *William*, and *Nan*, a pet form of *Ann*. Again, it is not clear why these names in particular should have been adopted for this animal.

*Tom*, for a male cat, is of course the familiar personal name that is a short form of *Thomas*. The name has been used of other animals, such as *tomtit* for a bluetit, but its particular association with the cat owes much to an anonymous work entitled *The Life and Adventures of a Cat* published in 1760, in which the hero was a male cat named *Tom*, frequently referred to as "Tom the Cat," just as "Tybert the Catte" was in Caxton's version of *Reynard the Fox*. Like *Jack*, *Tom* also has its use as a general term for a human male, notably in *tomboy* (a word found earlier than *tomcat*).

The use of *robin* for the bird may well have been prompted by the alliteration of the personal name with *redbreast* (both the *r* and the *b*), just as KITTY alliterates with *cat* and POLLY with *parrot*. As always, it is the pet form of the name that is adopted for anthropomorphic use, in this case from *Robert*. As with other animal names, too, it first prefixed the standard word before superseding it, that is, the sequence went *redbreast* (first recorded in 1401), *Robin redbreast* (1450), *Robin (robin)* (1549). *Robin* as a popular personal name is also familiar in the human world from *Robin Hood* and *Robin Goodfellow*, former famous bandits in stories dating from the 14th century. It is the bird name, incidentally, that seems to have promoted *Robin* as a female personal name in recent times (compare similar names such as *Linnet*, *Mavis*, and *Merle*).

We can perhaps bracket together the names of the *jackdaw* and *magpie*, as they are somewhat similar in composition. (The birds themselves also both happen to be notorious thieves.) The former bird was originally known as the *daw*, although this name is recorded no earlier than the 15th century. It apparently acquired its *jack* prefix soon after. The latter bird is recorded as *pie* in the 13th century (the word came into English from the French), and took its *mag* some time before the early 17th century, when it is first found. Why the female name, and why this one? Although, as stated, *Mag* is a pet form of *Margaret*, the bird appears to have gained its nickname on French soil, as *Margot la pie*. This was then adopted by the English, initially as *maggot pie*, doubtless under the influence of *maggot* and (edible) *pie*. The French name is on record for the bird as early as the 14th century, and may in turn have come from some fable.

*Martin* as a bird name is said to have been given to the particular species on account of its migratory habits: it arrives in *March* and departs about the time of *Martinmas* (the feast-day of St. Martin, November 11). But this may simply be an attempt to explain the particular name, which is otherwise probably arbitrary. (Some writers have not very convincingly related the word to Latin *murus*, "wall.") It almost certainly came into English from French, where, as mentioned, the personal name was popular from St. Martin of Tours, and where it is found for the regular names of some birds, especially the hen harrier (*martin-chasseur*, literally "Martin the hunter") and kingfisher (*martin-pêcheur*, "Martin the fisher").

This brings us to the *dicky bird*. The personal name is a pet form of *Dick*, itself a short form of *Richard*. As a nickname, *dicky* is usually applied to a small bird, and the name seems to suggest something tiny or a quickly repeated action. Compare such words as *dainty*, *dinky* (although more recent, from dialect *dink*, "trim," "neatly dressed") and *tick*. This last word means both "recurrent tapping," like that of a bird's pecking, and "small creature." But such an origin may be overingenious, and the name may have been more obviously prompted by the sound of birds chirping, as it were "dicky-dicky-dicky." Compare PHILIP, above, as a name for the sparrow. *Kittiwake*, incidentally, as a bird name, does not derive from *Kitty* but simply represents the cry of the bird, a species of seagull. The same is true of *katydid* as a type of grasshopper native to North America. The association with the girl's name, however, was reinforced by Susan Coolidge's popular children's novel, *What Katy Did* (1872).

Mention of names for small animals is a reminder that ANTHONY exists as a term for the smallest pig in a litter. The word is sometimes found in the form *tantony* or *tantony pig*. The initial *t* here shows the origin of the expression in the name of *St. Anthony*, the patron saint of swineherds, to whom one piglet in each litter was at one time traditionally pledged.

There are, of course, some generic names for animals that are not derived from personal names. One of the best known is PUSS or PUSSY for a cat. It *is* a name, rather than a standard word, since it is used to call a cat, especially one whose actual name is unknown. It is found in a number of European languages, from Dutch *poes* to Gaelic *pus*, and represents either the spitting of a cat or, more probably, the meaningless utterance made when actually calling a cat, a form of voiced *ps*, itself perhaps intended to express endearment, as a sort of "kiss." This origin appears to be supported by the conventional word for summoning a cat in other languages. In Russian, for example, it actually is *kis-kis*. In non-English languages, however, the equivalent of *puss* remains a call-name, not a name for the cat itself. In fact PUSS as a name for a cat in English has now been virtually superseded by PUSSY, a pet form of it. (PUSS has also been used as a generic name for a hare, a fact which has led some to derive the name itself from *lepus*, the Latin word for this animal. It is a curious coincidence that the Romans used this word as a term of endearment, just as we use *Puss* today.) The popular concept of the cat as a female animal, whatever its gender, led to the use of *puss* and *pussy* as a playfully pejorative term for a girl or woman (typically as "sly puss"). Hence the more recent use (from the 19th century) of *pussy* in its coarse slang sense of "female genitals." The name, in its use for a cat, was popularized from the 18th century by English translations of Perrault's famous fairy stories, one of which was *le Chat botté*, literally "the booted cat," rendered in English as *Puss in Boots*.

Claims that PUSS derives from *Pasht*, a form of the name of the ancient Egyptian cat goddess *Bast* or *Bastet*, are purely speculative.

The French equivalent for PUSSY is MINOU, as a general affectionate name for a cat. But whereas PUSS or PUSSY has few if any variant forms, MINOU has several. The French novelist Pierre Loti said that he could never think of an adult female cat as anything but MOUMOUTTE, and of a kitten as anything but MIMI. Well-known individual cats with related names are MINON and MISTIGRIS (see Chapter 12, page 154). These names, and others like them, are directly related to French *mignon*, "dear," "darling," and collectively originate from a base element *mi-* expressing smallness, as also found in English *mini* and *minnow*. (The French critic Sainte-Beuve had a female cat named MIGNONNE.) MINET and its feminine form MINETTE likewise exist in French as a related name for a cat, and in regular colloquial speech similarly mean "darling." It is of note, too, that French *minette* has exactly the same coarse slang sense as *pussy* has in English.

Related names exist in other languages. The Italian equivalent of MINOU, for example, is MICIA, with MICINO meaning "kitten" and MICIO specifically "tomcat." (A cat named MICETTO was born in the Vatican and raised by Pope Leo XII in the 1820s. On the pope's death, the cat was adopted by the French writer and politician Chateaubriand.) Even German

MITZI expresses the "littleness," although as a personal name it is a pet form of *Maria*. And despite its quite different literal sense, the English cat name MITTENS may similarly be regarded as belonging in some way to this family. The name will be familiar to Beatrix Potter fans.

A former general name for a cat, especially an aging she-cat, was GRIMALKIN. It is first found in Shakespeare's *Macbeth* (1605), where it is used as the name of the demon spirit of one of the Three Witches: "I come, Gray-Malkin!" (I,i,9). (The spirit is in the form of a cat, and is later described as "brinded," i.e., brindled.) The name appears to derive from a combination of *gray* and *Malkin*. The latter word was at one time a personal name (a diminutive of *Matilda* or *Maud*) used as a general name for a poor woman. It was also the name for a specter or female demon: hence its use by Shakespeare.

Dogs have so far lacked a showing in this chapter. This is because there is no obvious generic name for a dog corresponding to PUSS (or PUSSY) for a cat. Dogs have traditionally come to be seen as individual, domesticated animals, each with its own individual name. Cats, on the other hand, frequently exist in a wild or feral state, and are not necessarily dependent for their existence on a human home. Put another way, the actual need to call a dog whose name is unknown will rarely arise. Most dogs accompany their owners when out. Cats, however, roam freely on their own, even when fully domesticated. Even so, there are some *near*-generic names for dogs, and they should be considered here.

Perhaps the most familiar is ROVER. This may have developed from the original sense of the word as "robber" as much as its later and more common sense of "wanderer," since dogs have a reputation as marauders. It was popularized by the British silent movie melodrama *Rescued by Rover* (1905), in which a dog of the name (really named BLAIR) frees a girl tied to a railroad track. It existed long before this, however, and is found as the name of a dog who is the subject of a poem by Jonathan Swift (1667–1745). SPOT is another common name for a dog, strictly speaking a spotted one, or one with a single prominent mark. TOWZER has also been popular in the past, especially as the name of a large dog used to bait bulls or bears. The name literally means that the dog is one that *touses*, that is, worries or pulls about. The word (and so the name) is related to both *tease* and *tousle*. Yet another standard dog name is FIDO, traditionally used by the media to refer to dogs in general. *The Times* of May 12, 1992, for example, has an article on pet care by Nicole Swengley headed "Groomed to cope with Fido." The name represents Italian *fido*, "faithful" (which happens to be exactly the same word as Latin *fido*, "I trust"). Dogs are legendary for their fidelity. (For a similar name, see TRAY in Chapter 12, page 163.)

A former fairly common general name for a pet dog was SHOCK, as in the poem by John Gay (1685–1732), *An Elegy on a Lap-dog:* "Shock's fate I mourn; poor *Shock* is now no more." The name was near-generic for a poodle, or at any rate for a dog with long shaggy hair (compare modern "shock-headed").

In modern times a familiar name for a dog has been BONZO. Here the precise source is known. The name was invented by the editor of the British magazine *The Sketch,* Bruce S. Ingram, and was introduced as the name of the comically shaped puppy that featured in the cartoons of G.E. Studdy, the first of which appeared in the issue of *The Sketch* for November 8, 1922. Where Ingram himself got the name is uncertain, though he may have been aware of the already existing Australian slang word *bonzer* meaning "good," "excellent," or have had *bonanza* at the back of his mind. More recently the name has been promoted (albeit in another context) by the British pop group *Bonzo* Dog Doo-Dah Band, formed in 1966. The name has been also applied to other animals, as in the American movie *Bedtime for Bonzo* (1951), where the name is that of a chimpanzee.

The *bunny rabbit* also deserves mention here, if only because *Bunny* is now found as a personal name or nickname, either as a pet form of *Bernice* or some similar name or in other cases of arbitrary origin. The rabbit's name, however, does not relate to this. It is a pet form of *bun* as a dialect name for the rabbit itself. Its origin is disputed, but it may be significant that the same word is used for the tail of a hare, especially in Scottish speech.

Finally, somewhere in the book space should be found for the ubiquitous *teddy bear*. The famous toy bear, often simply called a *teddy,* derives its name from *Teddy,* a pet form of *Theodore*. The specific reference is to the American president *Theodore* Roosevelt (1858–1919), who was well known as a hunter of bears. The toy bear first became popular in about 1907, and its name was promoted, if not actually created, by a famous comic poem, accompanied by cartoons, in the *New York Times* of January 7, 1906. The poem was itself prompted by Roosevelt's hunting expeditions, and it told of the adventures of two bears called "Teddy B" and "Teddy G." These names in turn were transferred to two real bears presented to Bronx Zoo that year, where they were also known as "Roosevelt bears." Later, the name was exploited commercially by toy dealers, who imported toy bears from Germany and gained wide sales with their "Roosevelt Bears" or "Teddy Bears," with the latter name soon prevailing. The instant fashion set by the bears remains as popular today as it was then.

# 3

# *Descriptive Names*

Many animals have descriptive names, relating to such attributes as their color, markings, size, distinctive body parts, voice, habits, or nature (friendliness, laziness, timidity, greediness, independence, etc.). Such names can be common and obvious, such as BLACKIE for a black cat, or more subtle and even punning, such as AN MEE TOO for a Pekinese who follows his owner everywhere. Names of this type are usually derived from standard words (*black, and me too*), although the name itself may not be original. That is to say, the person who names a black cat BLACKY will almost certainly not have devised the name from the word *black* but will have given the animal the existing name BLACKY. Even so, it is itself based on a standard word.

There is no doubt that color is one of the most popular and obvious attributes to suggest an animal's name, and such names are regularly used for just about any color that any animal can be. The usual formation is the word for the color with the diminutive or "pet" suffix -Y or -IE, giving such common names as BLACKY/BLACKIE, BLUEY, BROWNY/BROWNIE, WHITY/WHITEY/WHITIE, and, especially for horses, more specific shades and colorings as CHESTNUT, DAPPLE, ROAN (or ROANY), SORREL and STRAWBERRY. Here also belong such names as AMBER, CARROTY, CHERRY, CINDY (for a gray animal), COPPER, DARKY, EBONY, GINGER, GOLDIE, HAZEL, JET, RUSS (implying either *russet* or in some cases *rust-colored*), RUSTY, SANDY, SILVER, SNOWY, SOOTY, TAWNY, and TOFFEE, each denoting a recognizable color. SHERRY and WHISKY are quite commonly used as animal names, and can be applied to various shades of brown or tan. They additionally evoke a drink that itself has agreeable social associations.

Many animals have coats of contrasting colors, typically either black and white or brown and white. In such instances the name usually refers to the combination rather than the individual colors, and commonly occurs as SMUDGE, SPLODGE, SPOTTY, STREAKY, and the like. CANDY is a popular name for a striped animal, since it both describes the animal's coloring and suggests a "sweet" nature or appearance. Names that have a double reference like this are felt to be specially satisfying, particularly when one reference is obvious and one private. One owner named her white sow BLANCHE, partly for its color, but also partly because she had a cousin of this name. Another owner named a hamster SANDY, both for its light brown coat and for the road, *Sandy* Lane, in which he lived. A third named his golden retriever AMBER both for the color of its coat and from the fact that the dog was "always on the go." (To what extent such puns are devised *after* the giving of the name is uncertain. Some of them may not be original. But this does not prevent them from being enjoyably apt.)

Animals named for a person or place form a distinct category of descriptive names, and are considered as *link names* in Chapter 5, page 45.

Not surprisingly, many animal namers like to devise something more ingenious for a color name than simply a name based on the color itself. GOLDEN PIPPIN is thus a pony with a coat that is the color of this apple. TAMMY is the short name of TAMARISK, a red roan pony who is "in the pink" in summer. MISTY MOUNTAIN is a large white horse, while BRACKEN is a red roan pony with a coat the color of dead bracken. SHADOW is a dark gray horse. CASHER, a marmalade cat, is so named as his coat when a kitten was the color of a new penny. Those hearing his name, and unaware of its origin, frequently suppose it to be "Kasha," and of foreign origin. BIANCO is a white-coated horse. His name also implies that he is "sweet," as Martini or Cinzano *Bianco* is. The white Persian cat PARSLEY also has a name referring to his color, although more obliquely. He is "whiter than white," like the advertising slogan for Persil washing powder, and *persil* is the French for "parsley." There may even have been an additional subconscious association here between "Persian" and "Persil."

MOONSTONE is a light gray pony, CADBURY is a chocolate-point Siamese, named for the famous make of chocolate, COOKIE is a ginger cat, the color of a ginger cookie, COLA is a German shepherd whose dark red is the color of Coca-Cola, JAM is a cat with a black coat, the color of blackberry jam, FLAME is a red setter (whose breed name even suggests the rays of the setting sun), MUSHY is a mushroom-colored cat, MANGO is a cat with a coat the color of mango chutney, and SNUFF is a pale brown labrador, who also likes "snuffing around." JINJEE is a ginger cat, with a name spelled in an original fashion, as often happens in the world of name creation. SUKY is a bantam with dark brown feathers, the color of *Suchard* chocolate. CHAMY is a champagne-colored horse, and SHANDY a light-brown horse, the color of a shandygaff. BLANCO is a white Sealyham, with a name found for other animals of this color. MISS MATTY is a "matt" or black cat, with a name suggesting that of *Miss Matty*, the central character in Elizabeth Gaskell's novel *Cranford* (1851). DUFFY is a black Irish wolfhound, with a name deriving from Irish *dubh*, "black." OMAR is a black cat, so named because he is "very Egyptian looking." MAHOGANY is a horse with a more subtle name. It alludes not only to his brown coat, but also to the movie *Mahogany* (1975), which starred the brown-skinned singer Diana Ross.

PANDA is a black and white cat, while DOMINO is a horse with a coat also of these contrasting colors. BADGER is a tabby and white cat, and RUSTY is a cat who is ginger and white. The cat CINDY was so named, explains her owner, "because her perfect black and white markings include a smudge on her nose which made me think of Cinderella." Her name is similar to that of CINDERS, a silver tabby. CLOWN is a Welsh pony with a black coat but white face and legs. PENGUIN is a black cat with "a lovely white front and belly." DOTTIE is a pony whose white coat has black dots. Somewhat similar is the mainly white cat PIPPS, who has three sandy spots near his tail, like the "pips" or insignia worn on the shoulder by junior officers in the British army to indicate their rank, in this case that of a captain.

ATLAS is a skewbald, whose coat suggests a map of the world. His owner claims she could clearly pick out America, Africa, and the British Isles. BUMBLE is a guinea pig with a coat resembling the coloring of a bumble bee, while POLLY is a multicolored guinea pig. His name is short for POLYCHROME.

A black and white border collie is named GUINNESS. His owner found him abandoned in a refuse dump as a pup, and because his coloring suggested a glass of stout, with its dark body and white head, she originally named him HALF-PINT. As the dog grew, however, she changed this to his present name. (When the dog's owner was appointed the British RSPCA's first woman chief inspector, a photo of the two appeared in *The Times* of June 30, 1987, the dog's "livery" of black coat, white chest, and black head with central white streak coordinating exactly with his mistress's uniform: white blouse with black tie and shoulder flashes, black skirt, black stockings.)

TIGGY is a tortoiseshell and white cat, resembling a *tiger*. PUSSY WILLOW is a gray cat with silver gray paws, while FRENCHIE is a tricolor papillon, whose breed name evokes the *French* national tricolor flag. OLD GLORY is a rabbit with a lined and spotted coat, so he bears the alternate name of the "Stars and Stripes," the national flag of the United States. RAWSHARK is a spotted Dalmatian with a coat suggesting the ink-blots in a *Rorschach* intelligence test. The pony CHERRY does not have a cherry-colored coat but one marked like that of an Indian's horse in a western. His full name is thus short for CHEROKEE LAD. RAINBOW is the name of a colored mare, while MEASLES is a spotted pony. COBWEB is the name of a Shetland sheepdog (Sheltie) bitch whose coat looks as if it is festooned in spiders' webs. FLICK is a pony with a coat that looks as if it has been flicked with a paintbrush, while SPLASH, somewhat similarly, is a black horse with a dash of white on his underside. An even more artistic name is that of the black and white cat AUBRETIA, so named for the 19th-century illustrator *Aubrey* Beardsley, noted for his black and white drawings.

The black and white cat GINNY (also known as JENNY) has a name that originally began as WHISKY, with reference to the blend of *Black and White* whisky produced by James Buchanan. Her owners sought something more original, however, so proposed an alternative in HAIG, for another blend of whisky. This was then altered to THE GENERAL, abbreviated as GEN, for *General Haig*, the British World War I army commander. A veterinarian treating the cat in due course informed her owners that she was female, not male, as they had supposed, so GEN was finally modified to GINNY. This cat's naming history is a good example of the many adaptations and alterations that can occur, either through choice or from necessity. In this instance, as frequently, the animal ended up with a name that bore no ostensible relationship to its original name, yet had evolved perfectly logically.

Sometimes the color name relates specifically to a particular part of the animal's body, especially where it is conspicuous, such as the legs, tail or head. SOCKS or SOX is thus commonly found for a dark-colored animal with white legs, typically a cat or horse, while TIP is almost standard for a dog with a white-tipped tail. Individually named animals in this category include BOOTY, BOOTIE, TWINKLETOES and WELLINGTON, cats with white "boots" or paws, MINSTREL, a black poodle with white toes, JESTER, a white lamb with two black legs, PLUG, a white bulldog with a black tail, suggesting a plug in a sink, DICKY, a cat with a white shirt front, and OZZIE, a white cat with a black front, so named for Oswald Mosley, the British Blackshirt leader. Similar ingenuity gave the names of McGUIRK, a black cat with white legs, suggesting the Scotsman in the pair Sperry and McGuirk as which Bob Hope and Bing Crosby disguise themselves in the movie *Road to Utopia* (1945), and MALVOLIO, a mynah bird with "yellow stockings," like the character in Shakespeare's *Twelfth Night*. More commercial than literary is the name of the cat HOVIS, so called for "the two thin 'slices' of brown running across her cheeks," and thus suggesting the well-known make of brown bread. The cat KERNIE has a name that arose as a shortened anglicized spelling of his original French name, QUEUE

NOIRE, "black tail," whose color contrasted with his otherwise mainly white body.

An animal's head itself has several features (ears, eyes, mouth) that may prompt a name. SHINER is thus a dog with a black eye, SPECS is a black-masked lovebird, PINKY is a cat with a pink nose, SANTA is a cat with a white beard, SATAN is a German shepherd with a black face, TASH is a cat with a white mustache, and TOPHAT is a white cat with a black spot on his head. A foal with a large head in relation to his body was named CLANGER: he resembled the small creatures with big heads that appeared in the British children's television program *The Clangers* (1969). Another animal with a big head is the cat TAD, whose name comes from *tadpole*.

The size, shape, prominence, even absence, of part of an animal's body may also readily produce a name. PADS is thus the short name of PADDLEPAWS, a cat with large paws, CHIPPENDALE is a Sealyham with "cabriole" legs, POBBLE is a cat with toes missing (from the "Pobble who has no toes" in the nonsense rhyme by Edward Lear), TRIPOD is a cat with a missing leg, and POPEYE is an Afghan hound with "sailor trousers." A cat named QUIP was so called because it had seven toes on its back feet: the name is short for SESQUIPEDALIUS, from *sesquipedalian,* "having a foot and a half," a mock-learned pun (or, indeed, quip). Rather more obviously, HOMESPUN is a cat whose coat resembles a bundle of *homespun* tweed drying in the sun after it has been dyed. Mention should also be made here of OCTOPUS, one of two kittens, who between them had eight feet. The name is subtler than it looks, since *octopus* literally means "eight feet," from the Greek, while at the same time punningly suggesting PUSS.

CURLICUE is a mongrel with a curly tail, while KINKY is a pug with a tail of identical shape. CHAUCER is a cat with a long tail, the name arising by way of another literary pun. The cat BERTIE has a tail fractured in the middle. He was originally named simply BROKEN TAIL, but this proved a mouthful, so was abbreviated to B.T., which in turn became the standard personal name BERTIE. (Compare the evolution of GINNY's name above.) SQUIRREL is a cat with a bushy tail. CATKIN is a cat with a tail resembling the well-known flower cluster, which was itself so named for its similarity to a cat's tail. STUMPY is a cat born with no tail, like the Manx cat BELLEVUE, whose name refers to the anatomical feature prominently displayed instead.

NELLIE is a cat with one eye, so that the name is the pet form of NELSON, for the British admiral who lost an eye in 1794 during the Revolutionary Wars against France. Another cat who became blind in one eye was named MISTY, for her indistinct sight. Yet another was MONOCLE, for obvious reasons. WINKIE is a hamster with "eyes like periwinkles," according to his owner, while RANI is a cat with a "caste mark," like that of an Indian woman. SPOCK cannot be the only animal, in this case a cat, to have large pointed ears, suggesting those of the character Mr. Spock in the television series *Star Trek*, played by Leonard Nimoy. SURREY is a papillon with a "fringe on top," as in the popular song from the musical *Oklahoma!* HIPPO is a mongrel with a large mouth, and BEAKY a budgerigar with a deformed beak. PIG is the name of a dachshund with an overshot jaw, while PORKY is an Alsatian that similarly has a pig-like face. SNOOPS is a Shetland sheepdog with a long nose. (The comic strip dog SNOOPY undoubtedly did much to promote this particular name. See him in Chapter 12, page 161.) WOO-HOO is a cat with a face resembling that of an owl. A cat born hairless was named BRYNY, after the bald actor Yul Brynner. LACY is a cat with fur prominent round her ears, like a lace edging. ROSETTA is a guinea pig whose hair has curled into rosettes. WHISKY is a cat with long whiskers, while VANDYKE is a cat with a distinctive mustache and beard. BA-BA is a golden retriever who had long ears as a puppy, like those of *Babar* the Elephant in the children's picture books by Jean de Brunhoff. CORNFLAKE is a horse who at the time of his birth had two little ears sticking up like crisp cornflakes. SCREWY BLUEY is a gray gelding who has a "knock-kneed" appearance and gait. BEINE (rhyming with "Dinah") is a miniature dachshund with short legs. The name is German for "legs," as befits his breed.

Distinctive markings also generate non–color-based names. TOSH (or FRED TOSH) is a goldfish with a mark over his mouth like a mustache. The name can be taken as a fairly subtle blend of French *tache,* "mark," "stain," and *tache,* an abbreviation of *mustache.* QUEST is a horse with a blaze in the shape of a question mark. Similar is LIGHTNING, a horse with a zigzag marking. JACEY (or J.C.) is a dog with a cross-shaped mark on his back, while LUCKY is a black cat with a white horseshoe mark there. A horse was named LADY not only because she was a mare but because her face markings suggested the outline of a woman with her arms round her head. The cat ARIES, nicknamed AIRY-FAIRY, is so called for the "ram's horns" symbol for this sign of the zodiac which appears in white and black on the cat's face, a horn curving round each eye. Here, too, mention should be made of DIAMOND LIL, a cat with a white diamond on her chest. She was given the nickname by which the actress Mae West was known after she had played the main role in the vaudeville play of this title (1948), written by herself.

An animal's overall size, weight and appearance may equally suggest a name. WEED, for example, is a pony who was very thin when bought, as was TWIGGY, the latter named for the British fashion model, famous for her slim figure and thin legs. TINY was a small kitten when first acquired, and retained the name, despite later growing into a real fat cat. PINI is an attractive bitch, a "pin-up." BERTHA is a heavy-looking sheepdog, so was compared to "Big Bertha," the nickname given to the powerful German howitzers in World War I. Another plump puppy was named PUDDING, and this same name was given to a cat "because of her enormous appetite and tummy."

A strong Welsh mountain pony was named SAMSON, for the biblical hero, while perversely (as can happen with nicknames) HERCULES the hamster was so called because he was weak and feeble, and anything but Herculean. The cat CASSIUS was so called because, like the Shakespearean character of this name in *Julius Caesar*, he had a "lean and hungry look." A puppy who looked helpless when first found was called HELP. Later, his owner altered this to HEP or HEPS, short for HEPHZIBAH, since calling him ("Help! Help!") led to embarrassing misunderstandings. SIRIUS, a boxer, was so named because of his *serious* expression (maybe he was also a "dog star"?), and the punningly appropriate name MAGNIFICAT was given by a parson to his fine feline. GEMINI was so named not because he was one of a pair of twins, but because he was a miniature beagle, both a *gem* and *mini*. HAGGIS and TANKY are two plump ponies. The former has a Scottish owner, who gave the name shortly after Burns Night, when haggis is eaten. The latter's size and shape suggest the tank that is both the military vehicle and the water container.

Occasionally an animal may be named for an object of some kind. An obvious example is ROCKY the tortoise, who looks like a *rock*. The horse ROCKY, however, is named for his gray, rock-like color, as well as for the fact that his owner's father was a keen follower of the American boxer *Rocky* Marciano. More directly, the tortoise HAGGIS is named for his resemblance to the well-known Scottish dish just mentioned, boiled in its round, dark-colored bag. But in this subcategory there is also scope for originality, as in the case of TULIP, the cat believed by his owners to resemble this flower. A cat named PUFF was so called because her appearance suggested that of a powder *puff*. (The name also happens to suggest "Puss.")

Sometimes the name refers directly or indirectly to an animal's breed or species. PICKLES is a fairly common name for a mongrel dog, suggesting the familiar mixture of preserved foods. More individualistically, BITSER is also a mongrel, "bits o' this and bits o' that." His name is similar to that of the cat ITSY, who was also "itsy-bitsy" or of mixed origin. A Chinese pug was named MING, while a Scottish cairn terrier was called ROBBIE, for the great Scottish national poet, Robert ("Robbie") Burns. MALIK is a part–Arab horse, with a name that is the Arabic word for "king." The owners of the cat PERSEPHONE SMITH gave her a name that whimsically reflects her mixture of breeds, part *Persian* and part *Siamese*. A different sort of verbal play gave the name of BURY the hatchet fish, alluding to the familiar metaphor. TORTY is not a tortoise, but a tortoiseshell cat. SHELLEY is another cat of this breed. CATTY is not a cat at all, but a catfish. GREG is a Shetland sheepdog: the name is short for MCGREGOR, a typical Scottish name. A Scottie dog was named ANGUS on similar lines. POPPY is simply a puppy, with his name formed by assimilating a standard word to a recognizable name, as happened with SIRIUS just mentioned.

Less originally, an animal may be given a name based on the regular word (common noun) for it, so that a hamster is called HAMMY, a tortoise TORTY, a puppy PUPPY, a mouse MOUSIE, a rat RATTY, and so on. But dog and horse owners are usually (although not always!) more imaginative than to call their animals DOGGY or HORSEY, and cat owners mostly bestow a more original name than KITTY or PUSSY (see Chapter 2, page 13). Such names are typically children's creations.

TI PING is a shih tzu, so named from a Chinese fictional character, and SUZUKI is a Pekinese, named not for the Japanese motor company but for the maid in Puccini's opera *Madame Butterfly*. Like AN MEE TOO mentioned in Chapter 2, POO-CHEE is a dog given a mock Chinese name, in his case one implying that he was a special pooch. YARRAMAN, a horse, was so named from the town in Australia, itself said to be so called from the Aboriginal word for "horse." A Great Dane named CHRISTIAN was so called as a tribute to the Danish writer Hans Christian Andersen, while LINCOLN is a beagle of American origin. The dachshund SIMPSON has a name that puns on "DAKS-Simpson," the well-known brand of menswear.

Names can sometimes change to be more appropriate as an animal grows up. This happened with a part-Siamese cat, who was initially called TINKERBELL. To reflect the oriental side of her breed, her name gradually changed through TINKER and TINKI-TONK to TON-KEE. The name of the cat PERCY BLOGGS indicates that he was not only a *pussy* (Percy) but that he was different from other cats, so was given a surname (Bloggs). PUSHKIN, rather similarly, is a name also frequently given to a cat, with punning reference to the Russian poet. For the same reason, some cats are called ALEXANDER, with the poet's first name. GINKY was the name given to a *guinea* pig, while on the same lines GERBELLA is a *gerbil*. Rather more originally, another guinea pig was called FIVER from the fact that it was valuable, as if worth five guineas. WINDY, a Lakeland terrier, was rather more obviously named for *Windermere,* the largest lake in England's Lake District. A German bulldog bitch was named ANNA, as her owners regarded this as a typical German name.

## 3. Descriptive Names

The actual sex of the animal may prompt a name. BUTCH, for example, is a common name for a male cat or rabbit. A female cat who looked more like a tom was given the compromise name of HE-SHE. It happens sometimes that a pet is believed to be male but turns out to be female, or vice versa. This can result in a change of name to reflect the true gender. Thus a budgerigar named JOEY was renamed JOSIE, a hamster HOMER became HOMERA, and a cat called DICK was transmogrified into MRS. DICK. When a male budgerigar called BILLY BOY turned out to be female, he (i.e., she) simply became BILLY GIRL. PADDY and MURPHY were two kittens with Irish names. When they were found to be female, however, they were renamed respectively as PERKY and CHOLMONDELEY. (The latter name is not as formidable as it looks, and is a surname traditionally pronounced "Chumly.") Such a name change is not obligatory, of course, and a cat named JIMMY retained the name when it was discovered that he was female, not male. Similarly, the black toy poodle CHARLEY-GIRL, found wounded in the street in 1986 by the television personality Katie Boyle, was originally named CHARLEY, as she was thought to be a male. When it was found that she was a bitch, her owner decided to keep her name because she "didn't like to add to her other traumas."

An animal's movement or gait may well suggest a suitable name. A lively pony was thus named FROLIC, while SWISHER is a horse with an active tail. As always, puns abound, so that BELTER is a hamster who "belts around," while TARANTELLA is a horse who is always dancing about. STUMBLEBUM is a boxer puppy, and a horse that has a habit of wandering off is called ROMANY, from his gypsylike nature. CLOCKWORK is a pony with a regular, "mechanical" gait, while WOOSHAMS is a fast-running cat. BOBBITY BUMPKINS is a small mongrel who has to jump up in long grass to see where she is going. BASHA is a bulldog bitch who "bashes the furniture." She would have been called BASHER if male. Somewhat similar is the name of BULLDOZER, an elderly cat who was short-sighted, and so "bulldozed" his way through objects. LIGHTNING is a fast-swimming goldfish, and BUCKET is a basset hound who "buckets along." The black Labrador GALE is so named because as a puppy she left a trail of havoc when rushing round the house. TIGGY is a poodle who likes chasing her owner and playing *tig*. GORDON the cat is so named because he prances about as if dancing the old Scottish dance known as the Gay Gordons. The cat TWIZZLE was so named because as a kitten she twirled rapidly in circles. YO-YO is a cat who as a kitten jumped up and down on the drain-board. BUCCANEER is a horse that *bucks,* and GULLIVER is a hunting pony who "travels" a lot. PAVLOVA is a horse that "dances," like the famous ballerina, while SUGAR RAY is one that "boxes," like the well-known American middleweight champion, Sugar Ray Robinson. SEBASTIAN is a horse that is a good runner, like the British athlete Sebastian Coe. PUSSY CAT is a horse noted for his "cat jumps." (Compare the horse of the same name in Chapter 4, page 38, who came by his name for quite a different reason.)

The character or nature of an animal is a frequent source of a name. Thus the pony SATAN is not only black but "devilish." PATIENCE is a Welsh pony so named not because he is patient, however, but because patience is needed to deal with him. A pony originally named SHANDY for his color was renamed TRISTRAM SHANDY when his owner read Laurence Sterne's novel so titled and felt that her pony was like the hero, wild and unpredictable. (Sterne presumably named his hero from the English dialect word *shandy*, meaning precisely this.) MEL, as his name implies, is a sweet-natured mongrel, from the Latin word for "honey." LADY is a common name for a gentle or well-bred female pony. The borzoi MILDEW was given his contrivedly punning name because he "grew on" his owners. JOY, more obviously, is a Labrador that brings this quality to the family who owns her. Names of this kind can be generally commendatory or pleasant, simply because the animal arouses a feeling of affection in its owner. Thus one horse is named HONEY simply because his owner greets him every morning with the words, "Hello, Honey." The horse named BABOUSHKA was so called because her owner mistakenly believed this to be the Russian word for "sweetheart." When she discovered that it actually meant "grandmother," however, she did not change it, as both she and the horse had grown accustomed to it.

Some owners devise almost unreasonably complex names, delighting in some private chain of thought. A dog named ASHE was so called because he was stupid. How come? He was a *dope*, and *dope* is also a word meaning "drug," and a drug is *hashish*, known colloquially as *hash*, or in a Cockney pronunciation, *'ash*!

There seem to be almost as many "baddies" as "goodies" when it comes to names of this type. PECKY, for instance, is a cat who is a "peck of trouble," while DENNIS is a cat who (one easily guesses) is a menace. BUMBLE is a cat who is a "little B." BITCH, on the other hand, is not a dog but a fish that fights others in the tank. SPITFIRE is a cat notorious for his spitting and "swearing," and for his general aggressiveness. The parson who named his miscreant cat EDOM did so from an apt biblical quote: "Over Edom will I cast out my shoe" (Psalm 60.8). An animal's sexual behavior, as well as its social, may equally prompt a name. RUDI is a cat who "does rude things," and CASANOVA is a male pet swordfish who is "always on the chase." The cat JULIET was so named because she was a constant attraction to the "Romeos" of her neighborhood. ABBA, on the other

### 3. Descriptive Names 33

hand, was simply a near-tame starling who produced many offspring. The punning reference is to the biblical "Abba, Father" (Romans 8.15).

As always, a personal reference may lie behind a name of this type. A cat was named SINTHIA not for its own "sins" but for those of its owner, who had had an illicit relationship with a neighbor. Hearing the name, most folk naturally assumed that the spelling was the conventional one, *Cynthia*. The joke was a private one between the owner and her friend.

Animals who repeatedly run away or disappear may earn their names from this proclivity. NUDDER is a dog who was always missing as a puppy, when his mistress would wonder "Where's that nudder pups?" (i.e., "that other pup"). HOUDINI must be a name found for several pets who somehow managed to break out of their confines, like the famous escape artist. A cockatoo named STANLEY was so called because she often broke free from her aviary to become an "explorer," while DASTARDLY is a mouse notorious for his many daring escapes. SCOOT belongs here, too, as a cat who frequently "scooted." The cat CHESHIRE had the habit of appearing and disappearing, like the Cheshire Cat in Lewis Carroll's *Alice in Wonderland*. A kitten who did likewise was named MAGIC, as if constantly performing a conjuring trick.

Young animals frequently require house-training, and their endeavors in this respect may sometimes earn them a name. Here therefore belong PIDDLE the cat, PENNY the collie puppy, who "spent a penny" indoors, PUDDLES the dog, WILLY WET the puppy, and TIDDLY-POOS the cat. At least the budgerigar FLOWERPOT had a well-defined area of operations. One of the most original names in this rather uncomplimentary category was that of the cat PEEDICAT, derived from what his owner described as "a hilarious transliteration in one of Hugh Walpole's books for the Russian for 'fifty'." (Linguists know this is *pyatdesyat*, but the Cyrillic letter *c* was not transliterated *s*, as it should have been.) The cat PANSY-POO may be included here. The first part of her name refers to the markings on her coat, which resemble the outline of a pansy. The latter part, however, relates to her habit of breaking wind as a kitten, especially when being fussed or petted.

Puns of course are found in this subgroup as elsewhere. JOHN KEATS is a Scottish terrier who could not be house-trained as a puppy so was "one whose name is writ in water," as the poet has been more reverentially commemorated. The dog CARPENTER, too, was so named as he always "did little jobs about the house," "made a bolt for the door," and so on. Again, it seems likely that such puns are not always original.

Animals invariably love their food, and their eating habits (and greed) can furnish a name. GUTTER is thus a greedy blackbird, a garden regular, and BISCUITS is a horse who comes up at a trot whenever they are produced. GOBBLES is a tropical fish who "eats" the filter bubbles in the tank, while GUZZIE is a red boxer who *guzzles* her food. DUSTBIN is a dog always trying to tip the lid off the trash can to get at the scraps, and HOOVER is the dog who usually "cleans up" after MILDEW, mentioned above. PINCH is a guinea pig who steals food, and OLIVER is a cat who is always "asking for more," like the workhouse boy in Dickens's novel. FLOSS is a foal with a liking for candy floss. TEASPOON, however, has a more modest appetite. As a kitten she never ate more than a teaspoon of food. At least one good mouser is named DIANA, for the goddess of the hunt, while a cat who got his dinner straight out of the river was named IZAAK, for Izaak Walton, famous author of *The Compleat Angler*.

An animal's personal habits or mannerisms may suggest an apt name of almost any kind. TITO GOBBI is thus a dog that slobbers or "gobs." SCHUMANN is a dog who carries off shoes. Rather similar is CARRAWAY, a Labrador who is forever carrying things off into corners. Perhaps he is also dark-coated, like the seed. MOUSE is actually a timid cat, feeding at night only and squeaking rather than mewing. She is also small and gray. POSY is a nosy guinea pig, and another inquisitive animal is the pony NOSEY LAD, who according to his owner "stops and stares at every gateway on the road." SANDY the cat likes playing with sand, while WHISKY "eats her whiskers." SOLO is a "loner" cocker spaniel, and another pet who enjoys his own company is the black cat TODD. His name derives from British rhyming slang "on one's tod," meaning "on one's own," itself from *Tod Sloan*, at one time a well-known jockey. PLONK, by contrast, enjoys company. He is a cat who is always "plonking" onto his owner's lap. The cat BONK was so called not for his promiscuity but "because of his playful way of 'bonking' [hitting] with his paw at his owners when he thought it was time for them to get up in the morning." The cat LOVER BOY, too, simply has an innocent name denoting his demonstrative affection.

BEETHOVEN is a cat noted for his "moonlight sonatas," while the cat BIZZIE LIZZIE is not only constantly curious but hyperactive. The dog JESTER is a "crazy fool." SUNSHINE is a pony who likes eating and sleeping in the sun, and the cat HASHISH spends much time sleeping also, as if drugged. PUFF is a dog who puffs and pants when back home after a walk. PUFFY, on the other hand, is a cat whose fur puffs up when she is scared. FLAPPER is a budgerigar who "really is one." The name puns on *flapper* as a nickname for a flirtatious or "flighty" young woman of the 1920s: BONNINGTON (doubtless BONNY for short) is a cat who was always climbing curtains as a kitten. She was thus named for Chris *Bonnington*, the British mountaineer. JEMIMA PUDDLEDUCK is a dog who not only waddles like a duck but is an "indoor performer." (The name itself is straight from Beatrix Potter.) The cat CHOPIN is said to enjoy listening to music. More active and aggressive is BUTTIE the goat. A gerbil was named HARRIET because she is a "home ruler." The personal name is a feminine form of *Henry*, which actually has this literal sense.

The sound that an animal makes can also give its name. NODDY-TUG is a cat whose head nods when she is asleep, but who when stroked makes a "brup" noise, like a tugboat. YACKIE is a noisy cockatoo, and PINKY is a chaffinch whose song is "pink-pink-pink." The cat SQEAKY (*sic*) was so named for her feeble mew, while another cat was named PERCY because he *purrs* a lot. The cat HISSING SYD was so called for his spitting. The name is that of the snake character in the popular record "Captain Beaky" by the Australian actor Keith Michell, in the top ten in 1980. The tortoise MR. RUSTY creaks as he moves. WOOFE is a dog with a rather obvious name. DILLY is a guinea pig whose cough suggests the voice of the cartoon character Dilly the Duck. The horse EPI has the full name of EPSILON BAY. The latter half of the name refers to his color. The first word is the name of the Greek letter "E," chosen to represent his high-pitched whinny ("E-e-e-e-e!"). He was originally called EPISTLE, but this name produced sniggers, so was altered accordingly.

Descriptive names can thus take all forms, and relate to many aspects of an animal's life and existence. Some names are gently or amusingly critical, but most are positive and approving, like that of the Irish setter SCARLET, named not for the color of her coat but because she is bright and lively, like *Scarlett* O'Hara, the high-spirited heroine of Margaret Mitchell's famous novel *Gone with the Wind* (1936).

# 4

# *Incident Names*

Many animals are named as the result of an incident that occurred at or soon after the time of their birth, finding, or acquisition. The incident may directly involve the animal itself or simply have been a connected or coincidental occurrence. Thus a horse who put her foot through a basket soon after arriving in her new home is named FOLLYFOOT, and a puppy born at Christmas is called CAROL. What distinguishes an incident name from a generally descriptive name is that it relates to a single event, or at any rate a brief one. The occurrence may be repeated, it is true, but it is its first occasion that prompts the name. This means that some of the descriptive names in the previous chapter may have originated as incident names. The horse called BISCUITS may have been so named when he first came trotting up for some, or the dog called DUSTBIN when he first made his foray for food. But information given by their owners suggests that this was not the case, and that the names resulted in each case from a regular habit.

It is known, however, that the names of the animals in this chapter were all given as a result of a single incident. We begin with the "animal-centered" names, those resulting from something the animal did or the way it looked soon after being born, found, or adopted.

SUNNY JIM was a dog who leaped in the air on his way to his first home from the police station, where he had been "taken in" as a stray. His name derives from Sunny Jim, the energetic character created by the American manufacturers of Force breakfast cereal. ("High o'er the fence leaps Sunny Jim, 'Force' is the food that raises him.") JUMBO is a dog who, on his initial arrival in a household, struggled along with a wood log in his mouth, like an elephant. LITTLE MOUSE is also a dog, so named because he squeaked when out of sight as a newborn pup, and was actually thought to *be* a mouse. AH CHOO LUCKY is a Pekinese who sneezed soon after being born ("Ah choo!"), so was *lucky* to be alive. RALLY is another frail Peke who *rallied* round soon after being born. He later became a professional showdog as RALEIGH'S REVENGE. Less happily, the foal LONESOME lost her mother at her birth. The stray bitch CINDERELLA was found looking as if she had escaped from somewhere, like the girl in the fairy tale who ran off from the ball at midnight. Another stray was the cat WAIFER, not only a "waif" but "wafer-thin." Similar was ALIKAT, a black and white cat found living a semiwild existence, like an *alley cat,* in a shed at the bottom of his owner's neighbor's garden. STRABY was also a stray, referred to at first by his owner as "that poor old stray boy." The last two words of this gave the cat his name.

FUNNY is a cat who mewed loudly when first seen in the cats' home, but who after his adoption by a family never mewed again. The cat CHUFFY looked pleased with himself ("chuffed") after drinking his first plate of warm milk. MR. PYM is a stray cat who walked through the window of his owner's house one day. His name relates to the popular play *Mr. Pim Passes By* by the British writer A.A. Milne, with a slight upgrading of the spelling of the name. FIREFLY is a pony who flew round the field when his original purchasers attempted to catch him. Somewhat similar is WINDMILL, a pony who as a foal galloped round and round his new field. SHREDDY is a guinea pig whose first food in his new home was a Shreddie cereal biscuit. TIPSY is a hamster who staggered round the course of a race he was entered for. Another sporting event gave the name of MINUS, a guinea pig who lost a rosette awarded him. YAM SENG is also a guinea pig, but with a more complex name. It represents the Chinese toast *yang shen,* "bottoms up" (literally "good health"), as that was the position he was in when first seen. JONAH is a cat who as a kitten swam in circles in a tank until rescued. His name recalls the biblical story of *Jonah* and the whale. PUSSY CAT is a horse who caught a mouse in his stable. (Animal owners do not hesitate to name one species after another when appropriate!) SPICE is a Welsh pony who pulled a packet of spice out of a shopping basket, spilling it on the ground. TOPSY is a cat who when first acquired gave a mew "on top C," and who then "just growed," like the little black girl in *Uncle Tom's Cabin.* The cat REMBRANDT belongs to an artist, it is true. But his name equally refers to the "two escapades with a paint pot" that he had as a kitten. Another cat, COCO, was so named because when his owner went to collect him he was "clowning about," like *Coco* the Clown. The cat TISHY was so called because, when given her first home as a kitten, she crossed her front legs like the comic horse of the name in the cartoons by Tom Webster in the London *Daily Mail.*

Names that are not directly animal-centered are probably more common, however. One distinct category consists of names relating to the time or occasion of an animal's birth, as for CAROL mentioned above.

*4. Incident Names* 39

JANSTAR is a pony born in *January*, with a *star* on her forehead. SURPRISE is a pony born sooner than expected, while SECRET was born to a mare not thought to be in foal at all. SUNDAY is a pony born on this day, THOR is a dog born on *Thursday,* itself named for this Norse god, and FRIDAY the cat is likewise named for the day of her birth. MAY DAY is a pony with a similar name, although the suggestion is also of a dangerous or difficult birth, from the international distress signal. WINTER MOON is a pony born in a snowstorm at 10 a.m. CORRY is a British golden cocker spaniel born in the Coronation Year (1953). CRISPIN, also British, is a horse born on Derby Day and named for one of the runners in that famous race. CASPAR is a bay cob born near Christmas, and so named for one of the Three Wise Men. SAINT is a dog born on Christmas Eve. The dog JINGLES was also born at this time. His name relates to the popular Christmas song "Jingle Bells." The poodle MARTA (*sic*) was born on the day that Princess Marina of Greece died (1968). MAGDALENA, on the other hand, is an Alsatian (German shepherd) bitch born when the song "Eva Magdalena" was popular, and CARRIE ANN is a dog named similarly for "her" song of the day.

The black and white cat HECATE was born on Hallowe'en (October 31), so was given the name of the Greek goddess associated with the night, the protectress of witches. The name proved a mouthful to say, however, so is usually shortened (appropriately) to CATTY. SALOTE is a Scottish terrier born at the time of the visit of Queen *Salote* of Tonga to Britain (1953). The horse STORM was named for the weather conditions at the time of her birth. HALLE, another horse, was born on *Hallowe'en* (October 31). CASSIUS is a dog born on the day of a big *Cassius* Clay fight. It goes without saying that he is a boxer. The cat FOURPENCE was born on Maundy Thursday, the day when the British sovereign distributes special "Maundy money" (including a fourpenny piece). DR. ZHIVAGO is horse born on the day this famous movie was released (1965). The exact parentage of the foal TRINIDAD is uncertain, but he was so named because he was known to be the offspring of one of three possible fathers. PADDY is a cat born on St. *Patrick's* Day (March 17). (The name is also apt for an animal that "pads" about.)

More allusively the horse CHARLIE was born on the birthday of the owner's second cousin, whose middle name is *Charles*. Somewhat similar is the name of ERNIE, a horse born after the necessary stud fee had been paid by a timely Premium Bond win, produced by the computer ERNIE (Electronic Random Number Indicator Equipment). MASEY is a horse born in May. BOLERO is a horse born on the auspicious day (February 6, 1981) when the British ice dancers Torvill and Dean won universal acclaim with their performance to Ravel's orchestral piece of this name. One might not guess that the horse GODFREY was so named as he was born on *Good Friday*.

Names referring to the circumstances of an animal's acquisition, rather than birth, are equally varied, and may be given for an event associated either with the animal itself or with its owner.

LUCKY was a cat rescued from a canal by two boys. Another rescue gave the name of the kitten NEPTUNE, found struggling in the waters of a harbor by the Marine Police. BENDIX was a kitten that had found its way into a Bendix washing machine, and that was discovered only when the machine was switched on. SEFTON was a cat found to be injured, like the British army horse of this name, wounded by a terrorist bomb in London in 1982, but subsequently recovering, as did the cat. More original is the name of TU-TU, a cat wounded by a .22 bullet just before he was adopted.

LIDO, on the other hand, was a cat acquired shortly after his owners had returned from a holiday in Lido di Jesolo, Italy. Another dog named YATZI was so called after his owner had returned from Sweden, where he was assured by a colleague that this was the Swedish for "gin and tonic"(!). KALU is a roller canary whose owners had just returned from a cruise to Buenos Aires. The name was that of a Latin American dance tune played on board during the journey. (It must have been Percy Faith's "Kahlua," composed 1965.) BEAUTY the horse was obtained on an occasion that his 12-year-old owner described as "the most beautiful day" of her life. SUMMER QUEEN is a chestnut mare acquired on Midsummer Day (June 24), while two animals given as Christmas presents are CRACKER the cat and SPARKY the dog. Another SPARKY is a British budgerigar found on Guy Fawkes Night (November 5), when bonfires are lit and fireworks are let off.

FATE is the name of a kitten given his owner by a friend on a day when the two met by chance. MONTE CARLO is a dog given as a present on the day of the Monte Carlo Rally. The pony RAIN was bought on a day when it was raining, while the tabby cat CLOUDY was similarly named for the weather state at the time of his purchase. STRAVINSKY is a tortoise acquired soon after the death of the well-known composer (1971). His owner has another tortoise called HOLST, so named as his owner had just bought a record of this composer's orchestral suite *The Planets*. The black Labrador MONTY was bought after a blackberrying excursion, so was named for Field Marshal Montgomery, famous for his dark *beret*. The dog PASK was named after his owners had returned from an Easter visit to Sweden, the Swedish for "Easter" being *påsk*. The Pekinese NANKI POO was given his appropriately Chinese-looking name after his owners had been to see the Gilbert and Sullivan comic opera *The Mikado*, in which Nanki Poo is the Mikado's son. OCTO is a dog given as a present by a lady to her husband on his birthday in October. MERRY was a boxer bought at Christmas, while the cat BOXER was a present given on Boxing Day (December 26). The terrier DAVID was got on St. David's Day (March 1), while the cat LUKE arrived on his owner's doorstep on St. Luke's Day (October 18). (The name also happens to suggest "lucky.") TAROT the cat was so named because he came to his owners the day after they had been at a Hallowe'en party (on October 31); tarot cards are used for fortune-telling, a traditional activity at this festival.

CHIANG is a chow named for (and with the permission of) Madame Chiang Kai-shek, after her owner had met the distinguished Chinese president's wife at the Chinese Embassy in Washington, D.C. PIE, more originally, is a Yorkshire terrier abandoned on the forecourt of a service station by an American family who were returning to the States after a visit to England. The garage owner's daughter, who had just begun to study algebra at school, commented that the dog was "the unknown quantity, $\pi r^2$" (actually the formula for calculating the area of a circle). More intimately, the puppy SHERRY got her name after her owner had been drinking the wine in question and had found it difficult to carry the dog upstairs. The stray dog OBEE was so named after his owners were taking him home from the police station and passed a Salvation Army band playing "O Be Joyful." WREN is a corgi brought to some kennels by a man who said that her name was something like "Rin." Her owners modified it to a recognizable word accordingly. Spoken words also gave the name of the fox terrier NUTTY. When the dog's new owners collected him as a stray from the police station one of the officers commented, "You're lovely, but you're nutty."

PROMISE is a pony whose owner's father had said "What a promise!" when her mother had vowed not to buy a pony at a sale, but did so. In like vein, HONESTY is a pony bought by a wife whose husband had not wished her to do so. She therefore felt obliged to confess: "Honesty is the best policy." Very similarly, FOLLY is a pony whose purchaser was encouraged by a friend to buy the animal, and who then confessed her *folly* to her husband. On the other hand, DICE is a pony whose purchase was arranged by telephone, so that it was something of a gamble. More straightforwardly, SCOUT is an Exmoor pony bought during a week when Boy Scouts were camping on this particular moor.

Some names are truly incidental. The pony BUBBLES was so named since at the time of his arrival a boy was blowing bubbles on the corner of the road. The pony ROXANNE was so called because she was born when the rock record of this name, by the British group Police, first entered the charts (1978). CLEMENTINE, another horse, was so named because she arrived just at the time when her new owner was finishing lunch with a clementine (a type of orange), and the foal PAPILLON acquired her name from the fact that when her owner was lying in bed, wondering what to call her, a butterfly flew in at the window. Another chance occasion gave the name of the Labrador CLOVA, born on the day that her owner's sister, out for a walk with her own dog, found a lucky four-leafed *clover*. More purposeful is the horse MUSTAVIM, whose name expresses the cry of his original owner on seeing him and suddenly wishing to buy him. Such chance occasions are in themselves often regarded as lucky, so that the name preserves their propitiousness.

Holidays are clearly an important enough family event to serve as a name for a pet, as for LIDO and KALU already mentioned. CWM, pronounced "Kwim," is a Scottish border collie named before the family who owned her went off on a holiday to Wales, where this word (meaning "valley") is found in many place-names. GILLY is a cat named after the mountain Sgurr nan *Gillean* on the Scottish island of Skye, where her owners had been holidaying. ASHBOCKER (*sic*) is a cat named after the English village of Ashbocking, Suffolk, through which her owners passed on their way to the coast. A cat named SANDRA was acquired by a family after a visit to *St. Andrews,* Scotland. If the cat had been male, his owners would have named him directly for the saint. The same people later named another cat CRAIG after they had stayed in Craigcrook Road. The dog DIGGER was so named before his mistress left for a holiday in Australia, where *digger* is a nickname for an Australian (from 19th-century gold miners), while KIWI is a cat belonging to a Canadian family who doubted whether it would be allowed to fly back to Canada with them from England: the kiwi is a flightless bird!

The actual cost of buying an animal in a pet shop or elsewhere may give its name. The cat DOLLAR is one so named, while the goldfish PENNY was won for a trifling sum at a fair. Another goldfish named ARTHUR, however, was bought in England for *half* a crown (2*s*. 6*d*. in predecimal currency).

The place where an animal was born, found or bought may equally furnish a name. The horse WILLOW was born under a willow tree, and the Welsh pony BRYN was born in Wales near a hill, the Welsh word for which is *bryn*. Another Welsh horse is BREC, born in Brecon. The two British horses DEVIL'S JUMP and WASHAWAY were born near these places in Cornwall. TIMOLIN is a horse born in the Irish town of this name. The dog BILBO was found abandoned in a bag as a pup, so was named after Bilbo Baggins, the hero of J.R.R. Tolkien's famous tale *The Hobbit* (1937). The puppy COP was given a home from a police station dog pound. The cat PUFF PUFF was found on a railroad station. CARO is a cat found abandoned on the wall of a *car* park, while OMNI is a cat found on a bus. (The former name could also be interpreted as Spanish or Italian *caro*, "dear.") LITTLEHAMPTON is a cat bought in this Sussex town. WHISKY is a cat who came from a pub. BLUEY came from a hotel called the "True Blue." BENEDICTA is a cat from a nunnery, where one of the sisters was so called. On similar lines, THEO is a cat who came from a *theo*logical college. UKEY is an American cat who came from the U.K. JASON is a dog found tethered as a stray outside "Jason's Cafe." The cat REMUS, like BENEDICTA, came from a convent. His name is short for OREMUS, Latin for "Let us pray." NICHOLLS is a cat from the farm of this name, and the cat CURLEWS came from a house that was so called. The cat BROOKWOOD was named for the hospital where he was found, living as a stray in its grounds. Similar to this is the cat EMI, adopted when living a precarious existence as a stray in the carpark of the company EMI (Electrical and Musical Industries).

The name of a street where an animal was found or had its home may also be adopted for it. ROBYN is a cairn terrier from Robyn's Way, while the cat HERM was found in Hermiston Avenue. The mongrel CHUA was found in Chua Beng Heng Road, Singapore. BUNBURY is a cat who was not only found in Bunbury Road but who is also notorious for his "bunburying," or habit of disappearing on private visits. (The term comes from Oscar Wilde's play *The Importance of Being Earnest*, in which one of the characters invents an imaginary sick friend in the country, Bunbury, whom he "visits" whenever he wishes to escape an engagement in town.) The cat DERRINGHAM was adopted from a feral colony living by a road of this name. He is known for short as DERRY or DERRY D. More allusively, the cat LUCY was named for the "Lucy" poems by William Wordsworth, since she and her owner live in Wordsworth Walk. Although not named for a street, the cat PORCHY really belongs here. When found as a stray in 1980, he was at first given a temporary home by his owners in a cardboard box in the *porch* of their home. As the weather turned colder, however, he was allowed inside the house, and took up permanent residence.

It is sometimes considered unlucky to rename an animal. Yet some renamed cats have already been mentioned, and the following horses were all renamed with incident names without any apparent ill consequences. GOLD DUST, so named for her value, was renamed CHARMER when bought, from the fact that her owner's mother had to sell her *charm* bracelet to raise the necessary cash. LEA likewise was originally LIZETTA, but gained her new name when her owner had a sore throat and could not speak the longer name properly. And the presidential REAGAN was rechristened BEAN when his new teenage owner found that the most successful command to make him respond was the mildly abusive, "Come here, you wally bean!"

Two final examples of an incident name will serve to illustrate the type further. The first is that of the cat OCKY. The name is short for OCTAVIUS, and was given him as he was the eighth cat to become a member of his owners' family. The second, more poignant, involves the cat SOS. On the very night of the death of one of two companion horses, the cat came to the stable to curl up on the back of the forlorn remaining horse, as if summoned by radio distress signal. He was a cottage pet from just up the road, and stayed a fortnight, after which he disappeared as suddenly as he had come. His particular incident name was really a nickname, given by the household who owned the horses, and his true name was not known. The story is told in Sir William Beach Thomas' memoirs *The Way of a Dog* (1948).

# 5

# *Link Names*

Many animals are given a name that already exists in its own right, whether as the name of a person, a place, or another animal. Such names may be thought of as "link names," since they *link* or associate the animal with the original name or its bearer. There have been examples of link names in the preceding chapters. Among them are NELSON the cat blind in one eye, named for the famous English admiral, the cat MR. PYM, named for the character in a play, the cat BRYNY, named for the actor Yul Brynner, and the horse TIMOLIN, named for his Irish birthplace.

The most common kind of link name is one adopted from that of a person, who in turn may be someone in the public eye or simply a family member or friend. The person, too, may be real (NELSON) or fictional (MR. PYM).

In a number of cases the personal name adopted is simply a general one, chosen to typify a particular characteristic. For example, an aristocratic-looking cat was given the "superior" name CARRUTHERS, while a similarly seeming corgi was called PARKER, this being a typical name for a butler. A "domineering" cat was given the name WINIFRED, which her owners felt itself suggested this attribute. MATILDA, on similar lines, is a haughty-looking cat. A sleek and elegant-looking pigeon was named HERBERT, while his more "down-market" companion was called CHARLEY. CLAUDE is a "snooty-looking" cat. (The name also, of course, punningly suggests *clawed*, so is frequently given to cats for this reason alone.) The tortoise SIDENEY was deliberately so named since his owners felt that he earned this "special" spelling, distinct from the regular *Sidney* or *Sydney*. The cat PRUDENCE is actually male, but was thought to be "full of feminine wiles," so was named accordingly. CHARLIE BOY is a "cheeky" dog, given a name to match. HUDSON HAMMY is a "superior" hamster, however, given a name associated with that of a well-groomed butler. (The name was popularly promoted by the fictional butler played by Gordon Jackson in the television series *Upstairs, Downstairs*, screened in Britain in the first half of the 1970s.) The timid cat BRUCE was given a name that his owner felt to be "strong" and "manly," one that would "bolster his courage." The cat ELVIRA had a shaky start to her life, so was given a "posh" name to boost her morale and self-esteem (really the owner's, of course), accordingly. The cat FANNY was so called because her owner felt the name was expressive of the animal's "downtrodden" state when she was found, and that the name itself had associations of poverty and wretchedness. GEORGE, a cat in an adult training center, was renamed PORGE (from *Georgie Porgie*) when a new director of the same name was appointed to the establishment.

A general name of this kind may be regarded as unsuitable, so is sometimes changed. A dirty, mangy cat was called CHARLEY when adopted. When properly treated and fattened up into an attractive tortoiseshell, however, she became EMILY. Her owner explains why: "The first major change was to be to her name. Charley was a ridiculous name for her and [...] she was definitely no charlie [stupid person]! So she became Emily. The name was chosen because it sounded pretty and delicate although was one which might have been better suited to the cat that I had pictured in my mind: a contented female tortoiseshell with clear golden eyes and a silky purr. Our Emily was really quite ugly."

Many personal names of this type are given on an apparently arbitrary basis, however, and owners asked to explain their choice of name frequently say little more than "It seemed to suit her" or "We just liked it." The instances below fall in this rather vague category. The owner's (or namer's) explanation for the choice of name is quoted verbatim in each case:

BEN, a Welsh collie, "For no particular reason"
BOBBY, a rabbit, "Bobby just suited her"
CHARLIE, a goldfish, "Because he had a mate called Freddy and I thought the two names went well together"
FRED, a goldfish, "Just liked Fred"
FRED, a goldfish, from owner's joking remark, "We will fry Fred for supper"
FRED, a rabbit, "No reason for his name"
FRED, a tortoise, "Pet name in the household"
GEORGE, a boxer, "Just a lovely name for a boxer"
JENNY, a duck, "Seems to suit a duck" (perhaps half suggested by *Jemina Puddleduck,* the duck in Beatrix Potter's children's stories)
JUDY, a cat, "Because Mum said so"
MATILDA, a praying mantis, "For no special reason" (but perhaps actually suggested by the name of the insect itself)

PETER, a cairn terrier, "I liked the name"
PRUNELLA, a white and brindle bulldog, "Prune-ella, lovely, isn't it?"
REUBEN, a sealpoint Siamese cat, "Somehow it just fitted perfectly"
ROSIE, a dog, "Because we liked the name"
SAM, a spider, "He knows his name well" (!)
SAMANTHA, a kitten, "Pretty kitten with a pretty name"
SHANE, a Labrador, "It suited his face"
SUE, a border collie, "Easy on the tongue"
TESS, a golden Labrador, "Couldn't think of a new name"
TILLY, a cat, "Just came"
TRIXIE, a dog, "I always wanted to call a dog Trixie"
ZUZIE, a mongrel, "Just a little more original than Susie"

A cat named HUGH was so called simply because the lady who owned him wanted to give him a more appropriate name than the peremptory YOU used by her husband to address the animal.

Traditional animal names are also given in this near-random way:

BRANDY, a Labrador, "Because we liked that name"
BUBBLE, a guinea pig, "The first name we thought of"
CORAL, a skewbald mare, "The name just suited her"
DIMPLE, a tabby cat, "Nice and feminine"
FIDO, an Airedale terrier, "Because my father, when a small boy, always wanted a dog called Fido"
FUDGE, a dog, "Looks like Fudge"
HAMLET, a hamster, "Because hamsters must have names beginning with H"
LASSIE, a bitch, "Because we liked it"
LASSIE, a collie, "All the others are called it"
PIPPIN, a Welsh collie, "Suits her well"
SNAGGLES, a cat, "Because we liked it"
TINKER, a cat, "General name for a scruffy animal"
TORTIE, a tortoise, "Couldn't think of anything else"

Sometimes an arbitrary name is given deliberately (so that it is strictly speaking not arbitrary!). This was the case for the cat JAPHET, whose owner was tired of explaining the pun behind the animal's original name, LAUDABLE PUS. This was an oblique reference to the work of Lister, whom the owner admired. (The phrase is a medical term for a healthy secretion of pus.) In other cases an apparently arbitrary name has enough thought behind it to take it just out of the genuinely arbitrary category. This is so for the Alsatian TOSCA, whose owner maintained that it was difficult to choose names for female animals because some of the best names are masculine, but who added that she liked *Tosca* "because it could be a masculine name."

An animal with a "regal" appearance may be named for a king or queen. JAMES was such a cat (his owner carefully explaining that the reference was to James II), and the same name was given to a rabbit "with a lordly air." (The name itself has aristocratic associations generally.) KING LOUIS is a lamb who seemed "royal" by comparison with his fellows. Also in this category may be included the crossbred Russian Blue/British Blue cat whose pedigree name is BLUE EMPEROR. His pet name is NIKKI, for Czar *Nicholas* II, the last "blue" (i.e., aristocratic) emperor of Russia.

The names of popular celebrities such as actors, musicians, and sports personalities are frequently drawn on for animal adoption, as are those of the fictional characters played by the actors. VALENTINE, for example, is a black cat named for the actor *Valentine* Dyall, who as "The Man in Black" narrated suspense stories on British radio in the 1940s. The black pug EARTHA, on the other hand, was named for the popular black singer *Eartha* Kitt. The cock-eyed cocker spaniel MARTY was so named for the bug-eyed comedian *Marty* Feldman, popular in Britain in the 1960s. POLLY was a corgi named for the fancied resemblance of her facial expression to that of Polly Elwes, the British television reporter on the *Tonight* program in the late 1950s. Some ten years earlier, the cat GUS was named for the popular American tennis player *Gussie* Moran, "Gorgeous Gussie." The cat had furry hindquarters, evoking the lace-trimmed panties sported by Moran on the staid courts of Wimbledon. A small, agile, and rather quaint cat adopted in the 1970s was named OLGA for the Russian gymnast *Olga* Korbut, then at her peak of world fame. In more recent times, the cat RUNSIE was named for the former Archbishop of Canterbury, Robert *Runcie*, with the *c* altered to *s* to make the name seem more feminine. The cat was given this name for her pleading or "prayerful" look when first adopted as a stray. (Her owner points out that the name is also suitable for an animal that "runs and sees," as a cat does when looking or hoping for food.)

The Pekinese puppy ENA appeared to have bags under her eyes, so was named for the character *Ena* Sharples played by Violet Carson in the popular British television soap *Coronation Street*, for whom this feature was also memorable. RIGSBY is an Old English sheepdog also named for a favorite television character, in this case the sneering landlord played by Leonard Rossiter in the 1970s British comedy series *Rising Damp*. The tribute arose from the fact that the markings on the dog's coat suggested the hole in the jacket regularly worn by Rigsby. Yet another television character, albeit a cartoon one, lies behind the name of CAPTAIN PUGWASH. He was a cat with a patch over one eye, like the pirate in the children's series named for him, first screened in 1957. Again, the cat PJ was named through his owner's admiration for Powlett Jones, the fictional schoolmaster in the television play *To Serve Them All My Days,* screened as a serial in 1980 from a novel by R.F. Delderfield.

Names of this type may equally be given to zoo animals rather than private pets. When in 1967 the eccentric English aristocrat Henry Thynne, 6th Marquess of Bath, opened Britain's first safari park at his Longleat, Wiltshire, home, he shocked many senior civil servants (but delighted the public) by naming five chimpanzees after prominent politicians of the day: HAROLD for prime minister Harold Wilson, GEORGE for foreign secretary George Brown, JIM for chancellor of the exchequer James ("Jim") Callaghan, BARBARA for transport minister Barbara Castle, and EDWARD for Conservative Party leader Edward Heath.

In some cases it may not be the animal who prompts the link with the person whose name is adopted, for example through a supposed physical resemblance, but the interest or taste of the animal's owner. Thus the owner of the Sealyham HUMPHREY enjoyed the playing of the jazz musician *Humphrey* Littleton, while BILLY the dog was named out of admiration for the British footballer *Billy* Bremner. The cat JO has the full name PUSSY-JOVA. Her owners were tennis fans, and keen followers of the Czech champion Vera Sukova, who first gained fame under her maiden name of Vera *Puzejova.* SACHA is a bulldog named as a result of his owner's admiration for the French singer *Sacha* Distel. Even a composite name may be adopted. The dog JEFFERSON was named for the American rock group *Jefferson* Airplane. His owners were admirers of the group, and named another dog FILLMORE for the *Fillmore* East auditorium, Greenwich Village, where they had enjoyed several rock concerts. (Strictly speaking this is the adoption of a place name.) The guinea pig GILBERT O'SULLIVAN was named not for the Irish-born pop singer but for the comic operettas of W.S. *Gilbert* and Arthur *Sullivan,* enjoyed by his owner. The cat TUPELO had an owner who was an Elvis Presley fan. He was therefore named for the Mississippi town where the famous rock and roll singer was born. More directly, the cat VANESSA was named for her owner's favorite actress, *Vanessa* Redgrave. The same owner had a cat named EDITH, since she in some way reminded him of the poet *Edith* Sitwell.

A witty verbal association of this kind gave the name of the rabbit STARSKY, who was surely not the only one to be so dubbed for the two Los Angeles cops of the television series *Starsky and Hutch,* first shown in 1976.

In a few cases an animal's name is given to match that of its owner.

This was the case for the cat TINKER, whose owner was a Miss *Taylor*.

As instanced by the cat NELSON, the admired person may be historical rather than contemporary. The Labrador GIDEON was owned by an apothecary, so was named for *Gideon* Delaune, royal apothecary to Queen Anne of Denmark. The Shetland sheepdog LAWRENCE was named for *Lawrence* of Arabia, whose life his owner was studying at the time of naming. Here, as always, a historical name may be adopted simply because it appeals in itself. Thus the tabby cat TABBYSHANKERS was loosely but not specifically named for the Russian World War II commander, General *Timoshenko*.

Names adopted from fictional characters may also be "animal-centered" or "owner-centered." In many cases they are the latter, but either way the names of animals are taken up as readily as those of humans. In some instances, too, it is a member of the owner's family who provides the link, not the actual owner. An example is the dog TOD, so named for the character *Toddie* in the popular children's book *Helen's Babies* by the American writer John Habberton, a favorite of the owner's father as a child. YOGI is a mongrel named for the owner's husband's favorite cartoon character, *Yogi* Bear. The Sealyham FLEUR was named for *Fleur* Forsyte, an admired character (played by the equally admirable Susan Hampshire) in the popular television adaptation of Galsworthy's *Forsyte Saga*, first screened in the late 1960s. (See below for a cat identically named.) The cat KIM was called after the young hero of Kipling's novel of the same name. FLASHER the Labrador was named for *Flashman*, the bully in Thomas Hughes' famous school story *Tom Brown's Schooldays*, a favorite of the owner's children. The dog PSMITH was named for the character in the stories by P.G. Wodehouse, his owners being dedicated fans of the writer. His near-namesake, the cat SMITH, was called after the cat of this name who appeared in the children's comic *Robin* in the late 1950s. The cat's owner kept the name for his successors.

The cat OLGA, mentioned above, produced kittens at the time her owners were watching the televised version of *War and Peace*. The arrivals were therefore named for characters in the novel: NATASHA, SONYA, PIERRE and NIKOLAI, so maintaining the Russian connection. SONYA then went to another family who already had a cat named SOAMES, for a leading character in Galsworthy's *Forsyte Saga*, also on television, and she was duly renamed FLEUR, for another important character in this work.

The dog SILAS, renamed SILESIA when found to be female, was named for Uncle *Silas*, one of the main characters in the stories by H.E. Bates, favorite reading of the dog's owner. The owner of the pony RUPERT was fond of the adventures of the cartoon character *Rupert* Bear when a child. TIGGY-WEE is a cat named for Mrs. *Tiggywinkle,* the hedgehog who was her owner's favorite character in the children's stories by Beatrix Potter. The cat TOPSY arrived when his owner was reading *Uncle Tom's Cabin,* in which the character of this name is a little black girl. (This last is also an incident name, therefore. Compare the black cat TOPSY on page 38.) More obliquely, the cat AMOS was so named because the owner's teenage daughter had a "crush" on *Amos* Burke, the police chief, played by Gene Barry, who was the central character in the American television series *Burke's Law,* screened in the 1960s.

The prevalence of character names from juvenile fiction here testifies to the role generally taken by pets as surrogate children in a family. Not all such literary references are juvenile, however. The cat NERISSA was so named to express her owner's love of Shakespeare. The named character is the waitingwoman and confidante of Portia in *The Merchant of Venice.* Perhaps the owner actually regarded the cat herself as a "waitingwoman and confidante." Many cats readily assume these roles. Some names of literary origin, too, are not for fictional characters but for a favorite author. The cat SNOW was so named when his owner had just finished the *Strangers and Brothers* sequence of novels by C.P. *Snow.* The name proved a talking point, as the cat was black, not white.

Animals may equally be named for some other animal, real or fictional. The dog SIMON, for example, was given the name of the cat hero on board the British gunboat H.M.S. *Amethyst* during the Chinese Civil War (1949). The mongrel MONTY was named for a red setter so called, while SHEBA is a dog named for a dog her owners met in a seaside hotel. One dog, RIBO, was named for his resemblance to *two* other dogs, respectively RICKIE and BONZO. Sometimes, too, an animal is named for its predecessor in a household, or for one of its parents. The cat HENDRIK was thus named for a cat HENRY who had just died, and whom he seemed to resemble. His pet names are RIK and WIKKET, the latter a mock-aristocratic variation on the former, with the affected pronunciation of *r* as *w* associated with the British aristocracy. A cat called POPPET was so named for her mother, POPPY. The name had an added personal appeal since her owner was regularly called "Poppet" by her boyfriend, subsequently her husband.

On the fictional side, BRAMBLE was a horse so named since he seemed to resemble the old nag in the British television comedy series *Steptoe and Son* (1964–73), in which Steptoe was played by Wilfrid *Bramble.* A dog ROVER was so called for his supposed resemblance to the dog of this name in the children's "Jack and Jill" reading books. A cat ROVER was also given the traditional dog's name, but came by it in a more complex way. His owner had seen *The Kidnappers* (1953), a movie set in a Nova Scotian village at the turn of the century. In this, a stern old man denies his young grandchildren the pet dog they want, so the two boys kidnap a baby, upon which the younger lad touchingly asks whether they can call it "Rover."

Only private information will reveal the source of an animal name adopted from that of a friend or family member. SALLY is a golden Labrador named for a girl at riding stables where her owner had learned to ride. The same owner had another golden Labrador named MARNIE for an outstanding girl rider in a pony club. Another Labrador was named BOGUS for her breeder, a Mrs. *Bogue*. At the same time, the name aptly described the dog's breed, which was not pure, but a quarter Irish setter.

The owner of the cat BISCOW named him for friends, while ELKE is a sheepdog named for a Swedish girl known to her owner. The mother of the owner of the pony DYLAN had herself ridden a pony belonging to a Mrs. *Dylan*, while STUTTERFORD is a cat named for a friend at the school his brother attended. The boy who owned CHRISTOPHER the cat named him for a friend of his, and the cat HANNAH was found by the owner's friend of this name. The pony DINTY PIPPIN was acquired with the help of a Mr. *Moore*, whose surname is popularly prefixed *Dinty* (as for the stew, itself named for the cartoon character in the comic strip "Bringing Up Father"), and a Mrs. *Pickin*, whom Mr. Moore erroneously referred to as "Mrs. *Pippin*." The border terrier ROBBIE is named for a family acquaintance, a Miss *Robson*, while the Great Dane CRICK is named for the local butcher. More daringly, EDWIN the longtailed scarlet shark fish is named for a neighbor described by his owners as "the local libertine."

As always, a name of this type may be combined with another type of name to form a composite. This happened with the cat MRS. NICOLA HOBBS. She was originally acquired by her husband and wife owners at Christmas and was thought to be male, so was accordingly named NICHOLAS (for St. Nicholas, or Santa Claus), while her brother was simply CHRISTMAS. When it was discovered that she was female, her name was duly feminized to NICOLA. It was then observed by the husband that the cat seemed to resemble the mother of a former school friend of his, a *Mrs. Hobbs*. She thus became MRS. NICOLA HOBBS.

This same cat's owners also devised link names for two seagulls that were "garden regulars." They were known respectively as STANLEY and MALVINA. The former name was inspired by the famous fictional seagull, JONATHAN LIVINGSTON SEAGULL (see Chapter 12, page 152), whose middle name evoked the historic meeting of 1871 between the explorer David Livingstone and H.M. *Stanley*. The latter name was suggested by *Las*

*Malvinas,* the Argentine name of the Falkland Islands, whose capital is *Stanley.* The two names thus link not only with a personal name and place-name, but with each other.

The owner of the dog JES, whose first name is *Jan,* named him for herself and *Des,* her boyfriend at the time she got the dog. (As she points out, she could have blended their names to give *Dan,* but decided in favor of something more original.) Another link with a former boyfriend lies behind the name of the Pekinese WIMOTHY TIGHT. In this case, however, the link is not with the person but the animal, for the friend's cat was called TIMOTHY WHITE, after a former British chainstore.

As with MRS. NICOLA HOBBS, a dual reference gave the name of HAMLYN, a school secretary's cat. The latter half of his name was adopted from that of his owner's aunt *Lyn,* short for *Lynette.* But the name as a whole refers to the "Pied Piper of *Hamelin*" of Robert Browning's poem, who was hired to rid the town of a plague of rats. The school was at one time similarly overrun with mice during the vacation, and as the "official" school cat had just died, the secretary bought a cat to remedy the situation, and named him accordingly.

Not surprisingly, some animals are named for the veterinarian who has treated them. Thus the bitch MARTINE was named for Dr. *Martin,* who saved her life when she was a puppy. (Her name is a feminine form of the surname.) MICHAEL, in turn, is a guinea pig named for a veterinarian's assistant.

Direct family involvement gave the name of TIMMY, a cat owned by Mr. and Mrs. *Tims.* The cat had earlier belonged to two boys, *Frank* and *Colin,* so had previously been known as FRANCO. The dog NELSON is so called as his owner's surname is *Hardy,* this being the name of the famous British admiral's flag captain. The hunter BRIGHAM is owned by a Mr. *Young,* thus punning on the name of Brigham Young, the American Mormon leader. A similar reference gave the name of the cat ISAAC, owned by a Mrs. *Newton.* The association with apples suggested the name BLOSSOM for the cat's sister. More straightforwardly, the rabbit NICOL was given a version of his owner's first name, *Nicky.*

Some surnames can provide a pleasant pun. A cat named TANGLE is owned by a Mrs. *Knott,* and a cairn terrier called DORA belonged to a Mrs. *Bool,* so that she was "adorabool."

A rare example of the reverse process, where a person is named for an animal, is that of owner of the horse GODFREY mentioned in Chapter 4, page 40. The lady concerned is named Wendy, who was so called for a mare WENDY, herself known thus from her place of origin, *Wenlock Edge,* the (windy) hill ridge in Shropshire, England.

Commercial names are sometimes pressed into service for animal use. A white and tortoiseshell cat was named CORTINA for her owner's white-colored car, a *Cortina* Estate. The cat's tortoiseshell and white sister was correspondingly named FIRENZA, for the darker-colored Vauxhall *Firenza* that was her owner's second car. Another cat, in another family, was named LOTUS for the owner's favorite make of automobile. The name is possibly more appropriate for a cat than a car in any case.

Occasionally a name refers to an owner's workplace or employer. The cat NEEB was thus named for the *N*orth *E*astern *E*lectricity *B*oard, for which the owner's son worked. More originally, the cat DINBA was named for his owner's former identity card number, which began with the letters *DNBA*.

In her book *The Cat's Whiskers* (see Bibliography), the British actress Beryl Reid describes how she came to name many of her cats for friends and fellow actors. They include the following (dates are those of appearances, not original writing):

BILLY, a ginger cat, for her friend *Billy* Chappell, who directed the show in which she appeared, *A Little Bit on the Side* (1983)
CLIVE, a long-haired ginger cat, for the actor *Clive* Francis, with whom she did a season in *The School for Scandal* (1984)
JENNY, a little tabby, for the actress *Jenny* Agutter, with whom she played at the National Theatre, London, in *Spring Awakening* (1974)
MURIEL for her veterinarian friend from the RSPCA, *Muriel* Carey
PATRICK, a handsome tabby, for the actor *Patrick* Cargill, with whom she played in *Blithe Spirit* (1970)
RONNIE for the diminutive comedian *Ronnie* Corbett, with whom she was on location making the movie *No Sex, Please — We're British* (1973) when she found the little black and white cat in a barn on a farm
SIR HARRY, a tabby, for her friend, popular singer *Sir Harry* Secombe

Other animals with "link" names belonging to Reid were:

ANDY, as a companion for another cat named FOOTY
ELLA (originally FOOTY; see ANDY), "because she had a very sexy walk," like the American singer *Ella* Fitzgerald
ELSIE, for a character's cat in a play starring Reid, *Born in the Gardens* (1979)
EMMA, a tiny tabby, "because she was like a little Jane Austen cat"
FURRY WEE (originally MIMOSA), initially FURY, for the famous "Wonder Horse," because of the cat's heavy footsteps, then by alteration to FURRY, then to FURRY WEE

GEORGIE GIRL, for the central character that she played in the movie *The Killing of Sister George* (1969) (based on the original stage version of 1965, in which Reid had also had this role), but doubtless also echoing the movie *Georgy Girl* (1966), and its title song

HAMISH, as a Gaelic name (the equivalent of *James*), for Reid's Scottish-born mother

KATH, a kitten found in a cemetery when on location filming *Entertaining Mr. Sloane* (1969), in which Reid played the character of this name

MOUNTBATTEN, a cat of Burmese appearance, for Earl *Mountbatten of Burma*

PARIS, for the figure in Roman mythology (whom Reid regards as "God of Love")

SID and RENE, two kittens, for characters played by Hugh Paddick and Reid in the television series *Wink to Me Only*

TUFNELL, for a character in a sketch performed by Irene Handl and Peter Sellers in *Swinging Sellers*

Not all Beryl Reid's cats were given link names of this type. One of her favorites was named DIMLY, not because he was stupid, but because he had just arrived in her home as a kitten when a friend, looking across a courtyard for Reid, said "I can just see her moving dimly." The cat thus has an incident name, so properly belongs to Chapter 4. He is admitted here, however, as an exceptional member of his owner's large feline family.

# 6

# Group Names

Many people own more than one animal, whether as pets or as "working" animals, for example on a farm. In such cases the animals may have their individual, distinctive names, or else names that interrelate in some way. Multiple names that have a common theme of some kind can be thought of as "group names," therefore.

The nature of such names frequently depends on the actual number of animals involved. A common number is two, in which case their names are obviously devised as pair or "twin" names. The animals themselves, incidentally, need not even necessarily be of the same species. And in some cases a "pair name" will need to be devised for a second animal who joins a household with a single animal in such a way that its name matches that of the existing resident.

BRANDY, for example, was a golden Labrador, so named for the color of his coat. His owners later acquired a kitten, whom they named SODA to match. Other pair names belong to HEROD and SHEBA, two bantam cockerels, GINGER and PICKLES, two kittens named for Beatrix Potter characters, MAC and ARONI (who came later), two puppies, ROMULUS and REMUS, two dogs, and CINDERELLA and ROCKEFELLA, another pair of dogs. The second name here punningly pairs with the first.

Sometimes, it seems, animal owners may give pairing names erroneously. TROTTY and PRINGY, for example, are two dachshunds supposedly named for the Charles Dickens characters Betsey *Trotwood* and "Miss *Pringle*." The writer's works, however, contain no character of the latter name. Other names are on safer ground. TALLY and BONT are two poodles born in a traffic hold-up in *Talybont,* Wales. SHIRT BUTTON and PEARL BUTTON are two puppies who chewed up a shirt, beginning with the buttons. MERRY KING and CAVALIER are two basset hounds, so named because their owner had been reading about Charles II, the "Merry Monarch," whose father, Charles I, had been supported in the Civil War by the Cavaliers. SHERRY and CYDER are two red-coated corgis, mother and daughter. The golden retrievers ALEXANDRA and ANGUS were born in the year that Princess *Alexandra* and Sir *Angus* Ogilvy were married (1963), while another royal wedding prompted the names of the retrievers MARGARET and ANTHONY, for the marriage (1960) of Princess *Margaret* and *Anthony* Armstrong-Jones, later Earl of Snowdon.

PUDDING and PIE are two dogs, mother and son, while TIGGER and WIGGER are two cats, jointly known as TWIGGER. These latter names are jingles, as are those of the two cats CHIFFY and CHUFFY. The two cats NICO and DEMUS are more meaningfully named. They both arrived by night, as *Nicodemus* did in the New Testament story when he visited Jesus (John 2,3). The splitting of a name or word like this into two is a fairly common device. Another example is the pair of Abyssinian cats named PEEKA and BOO.

LARES and PENATES are a dog and a cat, bearing the collective names of Roman gods who together were regarded as protectors of the house. More private are the names CARA and CHERRY, two cats named for the chalet, *Chacara,* where their owners spent their honeymoon. (As with the dog CARO in Chapter 4, the former name could also be understood as Spanish or Italian *cara,* "dear.") HANK and BRUCE are two budgerigars, named for *Hank* Marvin and *Bruce* Welch of The Shadows, the British pop group. WHISKY and SODA are two dogs more obviously named for their owner's favorite drink. SYDNEY and MELBOURNE are two kangaroo rats, named for well-known Australian cities, while MAIDA and VALE, a black Labrador bitch and her daughter, are named for the familiar London street.

The two names of a pair often combine very readily. PELL and MELL are two cats, as are HOLLY and IVY, suggesting the words of the Christmas carol. DERRY and TOMS are another pair, named for the former London department store, with the latter name obviously suitable for a male cat.

A perhaps apocryphal tale is told of two cats named ANCIENT and MODERN, "because they are both hims." But the hearer of this tale needs to be aware of the English churchbook *Hymns Ancient and Modern* to appreciate the pun. Better documented is the story of two oriental kittens, one very greedy, the other very sleepy, who were thus named CHEW and LIE, punning on the name of the Chinese Communist statesman who died in 1976, *Chou En-Lai.* ANDIE and FERGIE were a more recent pair of cats named in 1986 to commemorate the wedding of Prince *Andrew* to Sarah Ferguson, known popularly as *Fergie.* BERNARD and BOOTIFUL are another feline pair. They came from a turkey farm, so were known for the well-known commercial turkey farmer *Bernard* Matthews, who appears on television to advertise his turkeys as "bootiful" to eat. Slightly more subtle were the cats HINGE and BRACKETT (*sic*). They were two spayed (i.e., neutered) sisters named for the gentle television comedy act of Dr. Evadne *Hinge* and Dame Hilda *Bracket*, played by the British female impersonators George Logan and Patrick Fyffe. As nonfemales, they were also in a sense "neutered." Doubtless television also prompted the names of the two cats SPIKE and MILLIGAN, so called for the popular British comedian.

NICO and DEMUS, just mentioned, have a dual incident name. Another pair of cats with this type of name were FREYA and FRICKA, so called because on the evening they arrived, Wagner's opera *Das Rheingold*, which included these two characters, was being broadcast on television.

Sometimes the pairing or association of the names is only approximate. GEMINI and MELANIE are two dogs with such names. The only link is that both names have the same number of syllables as well as certain letters in common. ANGEL and DELIGHT, however, are two fantailed goldfish whose combined name refers to the proprietary brand of dessert. HECKLE and JECKLE are a pair of turtles, with apparently arbitrary rhyming names, although they could equally have been for the magpies of the animated cartoon. SCYLLA and CHARYBDIS are two donkeys, named for the pair of monsters in Greek mythology (and described by their owner as "little devils"). In similar classical vein, CASTOR and POLLUX are two kittens. Two pairs of ponies are ROCK and ROCK and TOPIC and PICNIC, the latter pair bearing proprietary names of chocolate bars.

THUNDER and LIGHTNING are sometimes found as a popular pair of descriptive names, referring to the animals' "stormy" nature or relating to their coloring. THUNDER may thus be a black horse, and LIGHTNING a white one, or else both animals may have streaky or zigzag markings. This particular pair of names is fairly widely found, as is its familiar German equivalent, DONNER and BLITZEN. The two cats POLLY and SUKIE are named for the nursery rhyme characters, with *Polly* putting the kettle on and *Sukie* taking it off again. Another pair of cats are called ALEXANDER and HERCULES, referring to the line from the anonymous song *The British Grenadiers*: "Some talk of Alexander, and some of Hercules." Cats frequently come in pairs, and further examples of pair names for them include ROSE and CAVALIER (together punningly evolved from the Strauss opera title, *Der Rosenkavalier*), BASS and TILLE (both born on July 14, the day of the storming of the Bastille, Paris), MILD and BITTER (for the traditionally paired English beers) and the famous couple KEITH and PROWSE, cats named for a leading London theatre ticket agency, whose advertising slogan was: "You want the best seats, we have them."

As always, the pairings of names may evolve on dubious linguistic grounds. TOSQUA and NINJA are a pair of tortoiseshell cats who "look Russian" and who were therefore given what their owner felt were Russian-sounding names. (Possibly they were actually suggested by *Toscanini,* however, the famous Italian orchestral conductor.) VIM and AJAX are two rabbits, one white, one black, named with the proprietary names of household cleansers. Their names seem to bear no clear relation to their coloring. The relevance is more apparent for the cats SALT and PEPPER, brother and sister, whose coats are respectively white and brown. WATTLE and DAUB are two ducks, one white, the other spotted. Their names refer to the familiar *wattle and daub* building material, comprising wickerwork (wattlework) plastered (daubed) with mud. WATTLE's name also happens to suggest *waddle,* the characteristic gait of a duck. The names would also be punningly suitable for a pair of turkeys, who have a *wattle* on their head or throat. Less enterprisingly, PUSSY ONE SPOT and PUSSY TWO SPOT are a pair of cats with the markings described. STING and OUCH are two bumblebee gobies, with names referring to their beelike "stings," while another pair of fish are the two kissing gouramis SLURP and SMACKER.

TOM and JERRY are commonly found as pairing names, for example for animals who chase each other like the familiar cartoon characters. (For the origins of this particular pair, see Chapter 12, page 163.) The owner of a pair of gerbils named them BOB and STAMP; when disturbed, one *bobs* up on his hind feet, the other *stamps* angrily. Two eccentric rabbits were named WOPSEY and WOOMSEY by their young owner. A pair of lovebirds were named PANIC and HYSTERIA. According to their owner, they both "bolt for the bedroom" when disturbed, but "Panic usually gives way to Hysteria."

It will be seen that many categories of the names considered in previous chapters are represented here. Several are descriptive. Others are associated by little more than sound. Sometimes categories blend, as they do for the pair of gerbils FRIENDLY and FIERCY. The names not only describe their individual natures (one can be held in the hand, the other cannot) but together have an association of sound and spelling, despite their contrasting meanings.

Groups of three animals can also have associated names. Three goslings were named RAMSBOTTOM, ENOCH and ME, from the names of characters in *The Happidrome,* a British radio comedy series of the 1940s. Its closing signature song had the lines: "We three in Happidrome, Working for the BBC, Ramsbottom and Enoch and me." Three Labradors with loosely linked operatic names are DIDO, PAGLIACCI and PAPAGENO, while more obviously associated are the three goldfish MEENY, MINY and MO, named from the children's counting-out rhyme. FERDINAND, NAPOLEON and JOSEPHINE are three hamsters, with contemporary historical names (*Ferdinand* III, Duke of Tuscany, sided against *Napoleon*, whose wife was *Josephine*). FLOPSY, MOPSY and COTTONTAIL are three rabbits named for their fictional equivalents in the stories by Beatrix Potter. The three mongrels PIP, SQUEAK and WILFRED were named for the three cartoon characters in a popular British comic strip running in the *Daily Mirror* to the 1950s. (They were actually a dog, a penguin, and a baby rabbit.) FREEMAN, HARDY and WILLIS are three kittens reared in a shoe box, and so appropriately named for the well-known British footwear company or its chain stores. More accessibly, MINI, MAXI and MIDI are three mice, the first of which was named for *Minnie* Mouse, wife of the famous Mickey. TIBBY, TABBY and TOBBY are three cats, with the names of the first and last randomly based on that of the second, a *tabby*. COS, SIN and SEC are the three kittens of a cat named TAN (for her color), the four names being standard abbreviations for trigonometrical functions (respectively *cosine, sine, secant* and *tangent*). The feline trio TAMMY, TIZZY and TUSKY, however, had little to unite them but their initial letter. JULIA, JULIANA and JULIET, on the other hand, are three cats with much more significant names. They were all born on March 15, otherwise the Ides of March, the day when *Julius* Caesar met his death.

When group names exceed three they are usually those of animals born in a single litter. Thus the four kittens BLYTH, BONNIE, GOOD and GAY were all "born on the Sabbath day," as in the nursery rhyme. The five kittens BORAGE, BAY, BERGAMOT, BASIL and BRYONY are all named for herbs as well as sharing an initial letter, denoting that they were born in the second litter. Their owner liked the theme so much that she retained it for the third litter, the names now beginning with C: CHAMOMILE (*sic*), CHERVIL, CHICORY, COMFREY and CORIANDER. Most such multiple names are found for pedigree animals, as it is in the world of show breeding that litters of this size will most frequently occur. They can be not only imaginative but meaningful. Thus the seven pugs PYRAMIDA, KING MAUSOLUS, DIANA, BABYLON, JUPITER, COLOSSUS and PHAROSA, named for the Seven Wonders of the World, were so called not only because they were seven in number but because they were born on the 17th day of the 7th month of 1967, took seven hours to be delivered, and belonged to an owner living in house number 57. (See Chapter 7 for further examples of this kind.)

For some animal groups or litters, the common theme can be worked in a more original way. In his autobiographical study *The Way of a Dog* (1948), the British journalist Sir William Beach Thomas, a noted writer on rural subjects, tells how his family had a terrier bitch who in 1877, the year of Queen Victoria's Silver Jubilee (fiftieth anniversary of her accession), gave birth to four puppies named respectively VIC, TORY, JUB and BILLY. Collectively, their four names are a form of link name.

Sometimes group names can be extended almost indefinitely, without being restricted to a particular number. The four cats SAPPHIRE, TOPAZE, JADE and JASPER have gem names that can readily be added to, while the four Dandie Dinmonts born in a car park, MINI, MAXI, MORRIS and MERCEDES, have automobile names that could also have been extended, although their namer restricts the choice by beginning each name with the same letter. The doctor who named his five Dandie Dinmonts PILLS, PHYSIC, TONIC, POULTICE and BORIC left himself greater scope to maneuver. Slightly more haphazard are the group names matching a cat called TWITCH (for his prominent whiskers). A second cat was named TWITCH-TWO and a third TWITCH-THREE. When this last was run over, his brother was named TWITCH-THREE-B.

Group names need not necessarily be given to animals born or acquired at more or less the same time. They can also be used for animals that succeed one another in ownership. Some owners, in fact, have a single name which they give to any animal that succeeds another, as a sort of "line of succession." An example of related successive names is that of the owner who had a dog WOOFER, so called as he was "fond of the sound of his own voice," and who gave his successors names with similar endings: GREGOR, DIGGER and COBBER.

When the number of animals in a household is large, however, and especially when they are (mostly) unrelated, they usually have individual names, without a common theme. A lady who had 14 cats thus named them as follows: BABY, HENRY, JOSEPHINE, LADYBIRD, LITTLE MAN, LIZZIE, LULU, NAPOLEON, PEACHY, POOKY, SWEEP, TIDDLY-POM, TIMMY, and TWIGGY. The only two in this group who were related in name were NAPOLEON (a female) and JOSEPHINE, her daughter. Distortion of historical genders and relationships is a liberty allowed animal namers!

It can equally happen that a group of related animals living together at any one time can come to have unrelated names. The names of the family of corgis owned by Queen Elizabeth II are a good example of this, and it is interesting to follow their evolution.

The present Queen Elizabeth had her first Welsh corgi presented to her in the 1930s when she was a small girl and simply Princess Elizabeth. The dog's kennel name was ROZAVEL GOLDEN EAGLE. However, his popular name within the royal family was DOOKIE, a nickname of obviously royal origin for a puppy who was "snooty" and "stuck-up" when being trained in the kennels, where he refused to eat from the same dish as the other pups. (The name evolved in stages: first THE DUKE, then DUKIE, then finally DOOKIE.) In 1938, when the Princess was 12, two more corgis were added to those that by now existed. They were born on Christmas Eve, so were appropriately called CRACKERS and CAROL. In 1944 the Princess' corgi SUSAN was born. Her pup SUGAR (a similar name) was born in 1949, and SUGAR's own pups, WHISKY and SHERRY, were given as presents to Prince Charles and Princess Anne in 1955, when the royal brother and sister were respectively aged seven and five. Elizabeth succeeded to the throne as Queen in 1952, and one of her favorite corgis came to be HEATHER, born in 1962. This dog was the mother of FOXY, TINY and BUSHY. FOXY then gave birth to BRUSH in 1969, and she in turn was the mother of Prince Edward's first corgis, JOLLY and SOX. A "hunting" theme may be discerned here, but only in the most general way. Subsequent royal corgis included PICKLES and TINKER, both born in 1971, and by 1986 the family of seven was as follows: DIAMOND, FABLE, KELPIE, MYTH, SMOKY, SHADOW and SPARKIE. In 1992 this had adjusted somewhat to: DIAMOND, FABLE, KELPIE, MYTH, PHAROS, PHOENIX and SPARKIE, names that display no common theme, although a family connection is indicated by names beginning with identical letters. The Queen chooses the names for her corgis herself, incidentally, and there is no "Mistress of the Dog" in the royal household to correspond to the historic royal post of "Master of the Horse."

The naming of related royal dogs with individual rather than thematic names is perhaps an established tradition, since a hundred years earlier, Queen Victoria's seven Pomeranians (then as fashionable in royal circles as Welsh corgis are now) were BEPPO, FLUFFY, GILDA, LULU, MINO and NINO. It is interesting, however, that five of the dogs, a German breed, have names of patently Italian origin. (The exceptions are the English FLUFFY and the German LULU.)

Queen Elizabeth's corgis have long been familiar to the British public, and although they may have individually distinct names, they have a common *nickname* in the media as "The Seven Dwarfs of Windsor." This name is smarter than it seems, since it not only puns on the title of Shakespeare's play *The Merry Wives of Windsor* but *corgi* literally means "dwarf dog." (See the breed names in Appendix I.) Moreover, the name implies that the dogs' mistress is "Snow White," who after all in the original fairy tale was a king's daughter, just as Queen Elizabeth is.

The converse of a collective name for a group of individual animals is a range of individual names for a single animal. Many such names may be regarded as nicknames, since they are additional to an animal's "basic" or main name. An example is the mongrel dog BAXTER, additionally known as BAXTER THE WALL (a whimsical expansion of the name), HOOVER (for his crumb-cleaning ability), PINCER TEEF (for his sharp bite), HITLER (for his single testicle, with reference to the popular bawdy song), GOEBALS (*sic*) (for his subsequent castration, ditto), NOEL (for his Christmastide birth), and HERCULE (in his role as "tec," like Agatha Christie's *Hercule* Poirot). Cats, in particular, frequently acquire a selection of different names, many of which are often variants on one another. Thus a cat with the main name TILLY had additional names TILLY WIMPOLE, WIMPOLE, WILSON, T. WILSON ESQ., M.P., TILLY WILLOW, TILLSOME PUSS, TWEEDIE and TILLY WILLIAMS. Her owner explained that WILLIAMS was a "surname" deliberately given to offset the name of TILLY-WILLY given the cat by others. TILLY's nine names (one for each life?) are beaten, however, by the 16 names of the cat SACHA. His battery of nicknames was as follows: ANIMAL, BUMPKIN, BUNGEY, BUSTIFER JONES, DIDYME, GAMIN, GOOFY, GUBBINS, GUGGLE-WUMPF, GUSTAVE, JO-JO, PLOUMEUR BODU, PUDDING, TWIDDELY-POOF and TWIDDELY-TOES. Most of these are affectionate "baby names," but ANIMAL was reserved for use after misbehavior, and BUSTIFER JONES alluded to his strength and "muscle power." It was no doubt directly suggested by BUSTOPHER JONES, the "cat about town" of the poem named for him in T.S. Eliot's *Old Possum's Book of Practical Cats* (1939). The curious French-looking PLOUMEUR BODU seems to suggest "plumper body," and *Bodu* may even be a typing error on the part of the informant for *Body*. This would tie in with PUDDING, denoting the "fat cat" that, as his owner reported, the animal became after his castration.

A special case of group naming is offered by the cat OLIVER RUPERT. His owners deliberately add a new name to his basic name every year, so that he is currently OLIVER RUPERT ALEXANDER BASIL RANDOLPH QUENTIN BARTHOLOMEW. He answers to any of these names individually when called. (But wouldn't a hungry cat answer to almost anything?) The reasons for these particular names is not known, but are on record for another cat similarly given additional names at intervals. His basic name is HUMPHREY, but his extended name is HUMPHREY ARCHIBALD MARMADUKE DANIEL "FIFI" ENTANCELIN SHPEEDY GONZALEZ JAMES LEWIS MARSHALL ELLIS. The added names originated as follows: ARCHIBALD for a friend who bawled at the cat one day when he misbehaved; MARMADUKE as he is a *marmalade* cat; DANIEL since he had some of the mannerisms of an earlier cat of this name; "FIFI" ENTANCELIN for his fast "cornering ability," like that of the 1930s racing driver of this name; SHPEEDY GONZALEZ similarly, for a racing driver of the 1950s; JAMES LEWIS for an elderly gentleman rudely awakened when the cat jumped on to his bed one night through an open window; MARSHALL ELLIS for an irascible neighbor whose ducklings the cat methodically caught and killed one day.

## 6. Group Names

In some instances, an animal can have a "group name" that is actually a single lengthy run of random or invented words. This is usually devised by a child as an enjoyable "pedigree" name. One such is that of the guinea pig POTOMAC VOLCANO SILVER BLACKBIRD BLACKY SIXPENCE (more regularly known as GUINSEA). More nonsensical were the names of two white mice kept in school by a young girl in the late 1920s, respectively INFRAGUMMIES INOKENUNNIES INFRADILGA INCLAYNANA MICKEY MOUSE BRITTAIN and ISABELLA BELLAMISSA SCHNOPSA BRITTAIN. The allure and originality of such names are attested by the fact that both were still vividly recalled by a schoolfriend of the owner some 70 years later.

# 7

# *Pedigree and Show Names*

"Professional" animals, those with pedigrees entered for shows, usually have two separate names. The first is a "pet name," used on the animal's home territory as the one by which it is called and spoken to. The second is a "professional name," a fairly formal affair, used for registering the animal with the appropriate association. The two types of names are not necessarily mutually exclusive, however, so that an animal's "pet name" may be incorporated in its "professional name." This situation usually applies to animals born not in a litter but singly, such as horses. Where animals are born in a litter, however, such as dogs and cats, they will usually begin life with a third name, given to distinguish them individually in the litter itself. This is in most cases incorporated in the animal's "professional name," when it is older. On the other hand, where young animals in a litter are sold off individually, as is often the case, the various purchasers will normally give the animal a new name.

Many examples of individual "pet names" have been quoted in the previous chapters. In the case of litters, however, as mentioned in Chapter 6, the young animals will normally be given group names, that is, names that distinguish them individually but unite them thematically or in some other way.

One system is to give the litter names that are associated with those of one or other (or both) of the animals' parents. A popular way of making the association is not only by theme but by letter. Let us begin with dog names, as the most generally familiar. (Cat names and horse names are on the whole more esoteric and exclusive.)

A golden cocker named SOLO, for example, produced a litter of eight whose names not only had a common link but all began with the letter S: SECOMBE, SEMPRINI, SYKES, SANTANA, SANGSTER, SATCHMO, STIRLING, SURTEES. These refer to celebrities who were familiar at the time of their birth, such as the singer Harry *Secombe,* the musician *Semprini,* the comic actor Eric *Sykes,* the jazz trumpeter Louis *"Satchmo"* Armstrong, and the racing drivers *Stirling* Moss and John *Surtees*. Another SOLO, a yellow Labrador, had pups whose names were associated not by theme but by letter only, each dog's name beginning with the first three letters of its mother's name: SOLILOQUY, SOLICITUDE, SOLICITOR, SOLITUDE, SOLSTAR, SOLITAIRE, SOLDIER BOY. A litter of Dalmatian puppies, whose mother was SHEBA, had names similarly linked only by initial letter: SEBASTIAN, SELINA, SPOTLIGHT, SCHUBERT, SHADRACH, SHINER, SEPTIMUS. When these seven puppies were sold, however, six were renamed by the respective purchasers, so that they became HECTOR, POLLY, RUPERT, HENRY, SHAMUS, and ANGUS. Only SHINER retained his original name. HECTOR and POLLY were in due course registered with the Kennel Club (see below), when they were further renamed with their "professional name," becoming respectively STARMEED SPECTRUM and STARMEED SATURN'S SIREN. STARMEED was the kennel prefix name.

Another system is one reflecting the owner's own personal interests. One music-lover gave her Irish setter's first litter the names SYMPHONY, ANDANTE, VIVACE, AMOROSO and BOLERO. A later litter had names paying tribute to the owner's daughter, who lived in France: ARC DE TRIOMPHE, MOULIN ROUGE, FOLIES BERGERES (*sic*), FLEUR DE LYS and MAMSELLE DE PARIS.

Sometimes a pedigree dog's name is mirrored not only in those of his siblings but in those of subsequent generations. Thus the descendants of the dog DUSTER and his wife MOP had names based on the "cleaning" theme that included BUCKET, SCRUBBER, SPIT, POLISH, SPONGE, FLANNEL, BRUSH, BROOM, WETTEX (a proprietary name), SQUEEGEE, CHAMOIS, SOAP, SWEEPER and BRILLO (another commercial name). In similar fashion two Jack Russells owned by a family with strong nautical connections were named TILLER and TRANSOM, and litters of their descendants included pups named BILGE, CLEAT, BLOCK, TACKLE, YAWL, KEEL, ROWLOCK, SPINNAKER, JIB, GYBE, SHACKLE and BAILER. The owners' yacht was itself named *Jenny Russell*, as the feminine equivalent (boats are traditionally "she") of *Jack Russell*.

Breeders of pedigree dogs frequently extend the thematic naming system to each generation of puppies in turn. A common device, again, is to use successive letters of the alphabet. Thus ARIELSTAR, already mentioned in Chapter 4 as an incident name, had a first litter with names beginning C: COMET, CONSTELLATION, CONVAIR, CLIPPER, CHIPMUNK, CORSAIR, CANBERRA, CERES, CUPID, all names of aircraft. (The dog is, after all, an Airedale!) Her second litter had names beginning E, continuing the "airy" theme but more poetically: ELEGANT, EROS, EAGLE, ECHO ESQUIRE, ELFIN, ERICA, ELAINE, EVENING STAR, ESTRALITA, ENCHANTRESS. (ESTRALITA suggests Spanish *estrella*, "star," and is also a near-anagram of ARIELSTAR.) The third litter contained only one puppy: FAIRY QUEEN. ARIELSTAR's son ECHO ESQUIRE then carried on the letters: GAZELLE, GILLIAN, GINA GAY KIM as the first litter, HARMONY, HERALD, HARLEQUIN, HAL as the second, and IVANHOE, IRVING, IMP as the third. Meanwhile ARIELSTAR's daughter FAIRY QUEEN herself produced only one puppy: JINDY LYNN. But this bitch in turn produced four litters, continuing the alphabet. The first three were: KNIGHTERRANT, KERRYGOLD, KANDIDA; LYNDALE KING, LAWMAN, LITTLEJOHN, LONE RANGER, LORRAINE, LUCINDA, LADYE GAYE, LUCKY CHARM; MARQUIN, MARSHALL (*sic*), MASCOT, MUSTANG, MELODY. JINDY LYNN's daughter KERRYGOLD (the trade name of a brand of butter, in fact) continued, producing five litters: NEMONIE, NINA, NERISSA, NECTAR, NUGGET; OLIVER, OBERON, OSWALD, ORIGEN, ORBITER, OSPREY, OLIVIA, OPHELIA, OLD GOLD, ORIEL, ORCHID BUD; QUARTERMASTER, QUIN, QUINOTE, QUINCE, QUILL, QUICKSILVER, QUAINT LADY, QUELIA, QUICKSTEP, QUALITY GIRL; REGAN, ROBIN, ROMEO, RANTAN, REDWING, RUSSET, RELCO, ROSALIND, RED CHERRY, ROSEBUD, RHAPSODY, ROWENA. Letter P had meanwhile been used for JINDY LYNN's fourth and last litter: PROSPERO, PALADIN, PEREGRIN, PORTIA, PERDITA, PHILOMEL. Letter S then followed for the litter of ARIELSTAR's granddaughter PERDITA: SAXON, SENTINEL, SCEPTRE, SYMPHONY, SERENADE, SAPPHIRE, SUNFLOWER, SONATA, STARSHINE, SAFFRON. KERRYGOLD then finally produced her fifth litter: TIMON, TAMORA, TITANIA, TAMARIND.

If an overall theme of any kind (apart from the alphabetical sequence) runs through these names, it appears to be generally Shakespearean, especially from letter N onwards. ARIELSTAR herself, after all, has a name suggesting a Shakespearean character, the "airy spirit" *Ariel* (in *The Tempest*), but the dog's name is primarily based on her breed of *Airedale*.

The official bodies responsible for the registration of pedigree dogs' names are the Kennel Club in Britain and the American Kennel Club in the United States, respectively. The Kennel Club (KC), based in London, is the older of the two, and was founded in 1873 with the chief objective of promoting the improvement of dogs and of regulating dog shows. The American Kennel Club (AKC) was formed in 1884 to take over from an earlier club set up in 1878, and is based in New York City. The New Zealand Kennel Club followed in 1886 and the Canadian Kennel Club two years later. The first dog registered with the AKC was an English setter named ADONIS. The first to be registered with the Canadian Kennel Club was also an English setter, FOREST FERN.

In 1880 the KC introduced a system of reserving the use of a name for a specific dog in the Stud Book in order to avoid a seemingly endless recurrence of dogs with identical names. BANG, BOB, JET, NETTLE, SPOT, and VIC were typical names of those years. The Stud Book itself, however, has been issued annually since the year of the club's foundation, while the Kennel Gazette, published monthly, made its first appearance in 1880. In 1900 over 10,000 dogs were registered with the KC; by 1980 the number had grown to around 200,000, with an annual registration figure of about 175,000.

The procedure for entering a dog's name with the KC is as follows. When a puppy is born, its owner (who normally is the owner of its mother, or dam) must complete a special form applying for registration, giving initially his or her own name and address and the name and address of the owner of the puppy's father or sire. The latter is required to sign a declaration confirming that the dam was indeed mated to the puppy's sire on a stated date and that the sire's name is already registered with the Kennel Club. The puppy's breeder signs a similar declaration to the effect that the names of both sire and dam are also already registered.

The breeder then suggests a "preferred" name for the puppy, and an alternate one, in case the first choice is unacceptable. The breeder is allowed to include a separate "affix" in the name, so long as the total length of the name does not exceed 24 letters. The "affix" itself is normally the name of the breeding establishment, or of the place where the puppy was born. STARMEED, quoted above, is such an "affix." In many cases the affix is actually a prefix, so that the dog's individual name is preceded by that of its place of origin. Examples are ALVERSTONE TIGGY and MERRISHAW ROSAMUND. Some breeders, however, prefer their kennel name as a suffix, following the dog's individual name, in which case it is usually preceded by OF. Examples of suffixed names are JAMES OF BURGERCROFT and SNOWDRIFT OF BEETOP.

It will be noticed that the breeder can be as inventive for the name of the kennel or breeding establishment as for that of the dog itself. This means that the private house where the dog was bred may actually have quite a different name, as in fact may the dog! On the other hand, most registered names are those actually in use for the animal, as is certainly the case for BABBACOMBE BUMBLE (the dog BUMBLE who comes from Babbacombe) and many others. A breeder is not obliged to include an affix in this way, but most breeders do.

Certain types of names are not accepted by the KC. They include the names of famous people (officially described as "notable persons"), names that are merely numbers in words or figures, and names that are simply a single word. Thus FLASH, as a single word, could not be registered, but FANNY FLASH could, as could (say) FLETTON FLASH. Nor will the KC admit single letters to be used as part of a name, or the breeder's initials, or even the breeder's own surname. There is thus nothing to prevent a breeder from using his or her own *first* name for the puppy, or the first name of any member of the breeder's family. It is therefore quite possible that dogs registered with names containing NANCY, JIM or KATE may be named for a breeder or family member. On the whole, however, most breeders prefer an established or obviously "canine" name, such as CRACKER, TRIXIE, or SANDY.

If the breeder chooses a name already registered with the KC, it can only be used so long as 10 years have passed since the original registration of the name and so long as the dog is of the same breed as its earlier namesake. This means that duplication of names is not permitted for dogs of different breeds.

A breeder may not propose a name for a puppy that is already part of the puppy's pedigree. This is not to avoid confusion over the name itself but because the pedigree name may well be an "affix" (as a kennel name) held by someone else, and this would automatically rule out the new name. Similarly, a breeder should not repeat a name when naming the different puppies of a litter, unless the name is registered as an "affix." For example, a puppy could be named TINKER so long as this name is already in use as a kennel name for other dogs owned by the breeder (say, as TINKER TRIXIE), but the breeder could not name two dogs TINKER if he or she used some other affix, or none at all.

If it turns out that the name is in some way unacceptable, the KC will then consider the breeder's proposed alternate name. If that is also unacceptable, one of two courses can be taken. The breeder can either be requested to select two more names, or the KC will itself select a name. Similarly, where two puppies have been given identical names, the KC will register the name for the first puppy but may well give their own name to the second.

Once a name has been accepted and registered, the dog's particulars will appear in the KC's Supplement, either as an original name, or on transfer to a new owner (with a new name), or as a change of name in itself. The latter occurrence is not common, however, and may simply involve the addition of an additional "affix." Thus the German shepherd GENTLE CASTRA OF REEVES was renamed with an additional suffix as GENTLE CASTRA OF REEVES OF RATHCONDEL, and a lowchen CLUNEEN MERRY MINUET was renamed as CLUNEEN MERRY MINUET OF LITTLECOURT. The latter dog thus has a registered name with both prefix and suffix. Both these changed names are more than 24 letters in length, but this is acceptable since they are not a first-time registration.

A typical Supplement entry might thus read as follows (the names of dogs and humans here are fictitious): "Malpas, Gregory of, d., Sept. 29, '91, br., Misses M.O., L.L., and Y. Malone; Solway Sam–Katie of Kismet; Mrs. I.M. Lovelee." This gives the following information. A puppy named GREGORY (probably actually called GREG or GREGGIE), at a kennel officially (but possibly not actually) named MALPAS, was whelped (born) on September 29, 1991, as a dog ("d."), not a bitch ("b."), in a litter resulting from the mating of the puppy's sire, SAM (kennel name SOLWAY), with the puppy's dam, KATIE (kennel name KISMET). The breeders of the puppy (and the owners of KATIE) were the three ladies, presumably sisters, surnamed Malone, while the owner of the puppy GREGORY is a Mrs. Lovelee. In cases where the dog also belongs to the breeder, its name will simply be repeated as the last item of the entry.

The AKC has laid down similar stipulations regarding permissible names. No name may contain Arabic or Roman numerals. Written numbers, on the other hand, are acceptable. The name must contain no more than 25 letters. Any words or abbreviations suggesting that the dog has earned an AKC title, such as *champ* or *winner,* are not permitted. Nor are degrading names, obscenities, or words in non-English alphabets. The following words must also not form part of a name: *bitch, dam, dog, female, kennel, male, sire.* The AKC allows 37 dogs within each breed to be given the same name. Obviously, the most popular names, such as LASSIE, PIERRE, SNOOPY, and SPOT, are already fully allotted. In certain instances, the AKC will accept the name of the breeder's choice, but will add a Roman numeral (despite the usual ban on this mentioned above) to distinguish it from other dogs of the same name.

An animal's "professional" or pedigree name will also appear in show results. The following names are those of the five dogs that were "Best of Breed" at Cruft's Dog Show in 1967. In their "titles," "SH.CH." stands for "Show Champion," while "AM." and "CAN.," more obviously, are "American" and "Canadian": SH.CH. CHESARA CHERVIL OF SEDORA; SH.CH. SUNREEF HARVEST GLOW; AM.CH. SALISYN'S MACDUFF; CAN.AM. DUAL CH. GRETCHEN V. GREIF; CH. BEHI'S CSIONS CSINY. In order, these five dogs are respectively a Sussex spaniel known at home as CHARLES, a Sussex spaniel known as MEG, a springer spaniel called DUFF, a German short-haired pointer with pet name SANDY, and a vizsla (a Hungarian breed of hunting dog) known at home as TRIXIE.

The interpretation of such official names can be a daunting business, as can the actual devising of the name, including that of the breeding establishment, in the first place. One owner chose MOUNTAINCREST as her kennel name since she lived at the foot of a mountain in Wales. Three of the dogs she bred (Shetland sheepdogs, despite their location) were officially named MOUNTAINCREST SPARKLER, MOUNTAINCREST WONDERBOY, and MOUNTAINCREST WELSH SPARK. Their pet names were respectively SPARKIE, TOPPER, and DEWI (the Welsh equivalent of *David*). An owner who lived in the Isle of Man bred a black-coated griffon Bruxellois whose professional registered name was ODDSOD OF DOO SCALLYWAG. Which is the dog's actual name here, and which the affix? The dog's individual name is SCALLYWAG, or SCALLY for short. His kennel name is DOO, the Manx word for "black," referring to his color. This was expanded to ODDSOD OF DOO as he was the "odd dog out" in his litter, with behavior running contrary to the general rule.

An animal's full registered name can thus tell an interesting story. Here is one more example. The boxer bitch registered as TINKER'S TRADE OF HARMAUR is actually known at home as AMY. HARMAUR is her kennel name. Her name TINKER'S TRADE relates punningly to her owners, respectively the breeder's daughter, Donna Wragg, and the breeder's friend, a Mrs. Potts. "Pots for rags" is a tinker's cry!

The rather unconventional or at any rate imaginative professional names cited here are nothing new. That such names have been devised and bestowed for a hundred years or more can be instanced by the following selection of registered names (with kennel affix omitted) included in the *Kennel Club Stud Book* for 1896. The names are those of winners in dog shows and field trials held the previous year. For ease of reference, the names are listed by breed, with explanatory comments as appropriate.

*Bloodhounds:* BOUNCER, BURSAR, WHAT'S WANTED
*Deerhounds:* SWIFTLY
*Greyhounds:* NO JOKE, QUITE A FOOL, HIGH AND INTERESTING, REAL JAM
*Foxhounds:* LADYBLUSH (by LUMEN out of LOVELY)
*Smooth-haired fox terriers:* LINGER LONGER LOO
*Wire-haired fox terriers:* GO BANG, MASTER BRISTLES (a dog), MISS BRISTLES (a bitch), THE PROGGINS (by PROVOST out of BRITON POP), SLAP BANG
*Pointers:* YZE (owned by a Mr. Anglois of Lyon, France)
*English setters:* KAISER, SWEEP THE GREEN
*Irish setters:* WHAT CARE I, JILL McSWINE (owned by Mrs. J. Swiney), NODNOL (a reversed spelling of *London*)
*Retrievers:* GOOD LAD
*Field spaniels:* SHOTOVER, SELAW (another reversal)
*Dachshunds:* KRON PRINZ, ERL KING, HONEY SUCKLE (a bitch), KNIRPS
*Newfoundlands:* HIS NIBS, PUDDING, RANDY

*Collies:* GOBANG, GRACE DARLING, LINGER LONGER LUCY
*Old English sheepdogs:* CHUM, RAGS AND TATTERS, SIR LOYNE
*Bulldogs:* ABRACADABRA (by CAT'S EYE out of PUSSY CAT), AUTOCRAT (by HIS LORDSHIP out of QUEEN ANNE), BABY BUGGINS, BOOM-DE-AY, THE CADGER, MONKEY BRAND, QUEER STREET, REV. DISMAL DOOM, AWFUL AFFLICTION, BUTTERCUP, CIGARETTE, ELECTRICITY, GROTESQUE, HEAVENLY PLEASURE, HOW NICE
*Bull terriers:* THE POP, UP TO DATE
*White English terriers:* WISE VIRGIN
*Skye terriers:* ANNIE ROONEY, YUM YUM
*Borzois:* KHAN (owner is "H.I.H. the Grand Duke Nicolas of Russia")
*Scottish terriers:* H.R.H., KILTS, WEE WEE
*Pomeranians:* BEN BOTHERIT, KOHINOOR, RAM JAM, MISS WIFFINS
*Maltese:* MELITA (the Roman name of Malta)
*Pugs:* COLONEL CHUBBY, VICAR OF LEEDS (owner lived in Leeds), POLLY WOLLY WOBBLES
*King Charles spaniels:* CHUMP, LE ROI, MERRY PRINCE, MUSIC, KOSMOS
*Poodles:* GENTLEMAN IN BLACK, THE GHOST, INKERMAN, MACARONI, PRINCE OF DARKNESS
*Toy terriers:* TOWER BRIDGE, CHEEKY
*Irish terriers:* COMMOTION, LADY GOLIGHTLY, LULU
*Mexican hairless:* THE HAIRY KING
*Basset hounds:* SHUFFLES, NOISY
*Airedale terriers:* THE BABE, BLOW, SECRET
*Irish wolfhounds:* WOLFGANG
*Whippets:* WINDY DICK, FLOREAT ETONA (owner lived in Eton; name is motto of Eton College)
*Schipperkes:* SPOOF
*Chows:* CHIN CHIN, YUM YUM
*Japanese spaniels:* STONEO BROKEO, THE MICROBE

If some of these names smack of the Victorian music hall, that is because they *are* of the Victorian music hall, in a period between the Crimean War and the Boer War, otherwise the second half of the 19th century. YUM YUM, for example, was almost certainly inspired by the character of this name (one of the "three little maids from school") in the Gilbert and Sullivan comic operetta *The Mikado* (1885), while BOOM-DE-AY echoes the popular song of American origin, "Ta-ra-ra-boom-de-ay," first sung in Britain in 1891 by Lottie Collins. Both LINGER LONGER LOO and LINGER LONGER LUCY owe their names to the popular song "Linger longer, Loo," sung at London's Gaiety Theatre in 1892 by Millie Hylton. This, after all, was the "Naughty Nineties"! INKERMAN, too, while being a name adopted from a Crimean War battle (1854), doubtless also referred to the dog's black coat. GRACE DARLING bears the name of the English national heroine, famous for her rescue (1838) of shipwrecked sailors. Some names reflect the dog's particular breed, such as the Scottish names for the Scottish terriers, the German names for the dachshunds, or the "Chinese" names for the chows. In many ways, too, the names are very similar to those of racehorses (see Chapter 8), especially where a dog's name reflects those of its sire ("by") and dam ("out of"). Individual names, as well, reveal pleasant puns: THE HAIRY KING is not only a pointedly contradictory name for its breed but also suggests "The Hurricane," as a fast runner.

So what of the cats?

The oldest cat club is the National Cat Club, founded in England in 1871. It was the cat fancy's first registering body, keeping a register of pedigree cats, granting championships, and issuing its first Stud Book in 1893. The club's annual show at Olympia, London, is still the biggest of its kind in the world. In 1910 the Governing Council of the Cat Fancy (GCCF) was set up to provide an "umbrella" organization for all the various British cat clubs and to coordinate matters of importance to the cat fancy. The GCCF also took over the registering and other functions of the National Cat Club, so that it is the body that now issues the regular Stud Book. There is now also the Cat Association of Great Britain (CA), founded in 1983 as an alternative to the GCCF and maintaining a register of pedigree, half pedigree, and nonpedigree cats.

Cat registration in the United States is somewhat fragmented, as there are at least nine registering bodies. The oldest is the American Cat Association (ACA), active since at least 1899, and operating in the southeast and southwest of the country. The largest American registry, however, is the Cat Fanciers' Association (CFA), whose annual yearbook is a major cat fancy publication. The most recent cat organization in the United States is The International Cat Association (TICA), now growing fast.

The GCCF actually based its regulations, including the registration of names, on those of the Kennel Club (see above), so that its principles of naming are similar.

Pedigree cat names consist of a prefix, common to all cats from a single breeding establishment, and an individual name, which may or may not, as for dogs, be the animal's "pet name." An example is the cat CORNWATER KIRSTY, where CORNWATER is the name of the breeding establishment and KIRSTY is her individual name.

A name may be submitted for registration without a prefix, but where this occurs the GCCF will supply an "administrative" prefix (see below).

The combination of individual name and prefix must not exceed 24 letters and must be two or three words in length. A name submitted *without* a prefix may not exceed 16 letters or two words in length.

The GCCF will not accept the following as a name or as part of a name: (1) the name of a living person, whether famous or not; (2) a registered trade name or business name; (3) a name containing a number, in figures or letters; (4) a name that repeats words previously registered by the same person, or under the same prefix; (5) a name containing punctuation marks such as hyphens or apostrophes. A name may be repeated, however, where the repetition is that of the recognized "color name" of the relevant breed. For example, BLUEBOY, as the individual name of a Blue Burmese, whether registered with a prefix or not (i.e., as SPRINGFIELD BLUEBOY, say, or as just BLUEBOY), can be used for another cat of this breed, just as WHITE LADY (or with prefix, say, DARIAN WHITE LADY), can be repeated for a second Longhair White. These restrictions broadly also apply to the prefixes themselves, except that a "color name" cannot be registered for a prefix, nor can a word that is similar to an existing prefix. The executive committee of the GCCF also reserves the right to reject any word that it regards as unsuitable. An example of the latter would be the prefix DIRTY, for this is derogatory and therefore undesirable for a self-respecting professional cat!

As is the case with the dogs, a cat's registered name may or may not relate to its "pet" name when at home. One owner chose the prefix name FLUFFENARD as a blend of FLUFFY, the name of her first cat, and *Enard Bay*, the location of her home on the northwest coast of Scotland. A Sorrel Abyssinian was acquired by his owner from a breeder whose prefix was LIBELLA. The breeder asked the cat's new owner to choose a name to match it, on condition it begin with the same letter. The new owner chose the name LABAJA under the impression that this was the name of a ruler of ancient Egypt (on the grounds that Abyssinian cats had descended from Egyptian cats). The cat eventually became a show-winner as CH. LIBELLA LABAJA. At home, however, he was simply RUSTY.

It will be observed that the GCCF's naming regulations are generally more stringent than those of the Kennel Club. Even so, the prohibition on using the name of a living person, for example, does not prevent many breeders selecting a name that fairly obviously suggests their own name, or that of a friend or family member. Evidence for such verbal manipulation is confirmed by a comparison between a cat's registered name and the name of its breeder, as recorded in the GCCF *Stud Book*. Examples include the following prefixes, extracted from the *Stud Book* of registrations from June 1, 1977, through May 31, 1980:

| Prefix name | Name of breeder(s) |
|---|---|
| BARLEE | Mrs. Barnes and Mrs. Leese |
| BERILLEON | Mrs. B. Lyon |
| BRIGIDEER | Mrs. G.E. Reed |
| BYRNETTS | Mrs. M. Byrne |
| CEEPAY | Mrs. C. Payne |
| CHALMI | Mrs. A. Imlach |
| DENEMS | Mr. and Mrs. D. Eames |
| DRURIES | Mr. and Mrs. Drury |
| ELSILRAC | Mrs. Carlisle |
| EMBEE | Mrs. M.G. Baxter |
| ESSAYCI | Miss S. Anderson-Caine |
| FRALLON | Mrs. A. Allon |
| JEANJEN | J. Jenson |
| KATHSHORT | Mrs. K. Short |
| LANDOSARLA | Mrs. S. Donaldson |
| LINDERN | Mrs. E. Mallinder |
| MINMOR | Mrs. M. Morcom |
| PEACHLYNN | Mrs. E.M. Champley |
| TAMBEMA | Mrs. P.E. Bateman |
| WILLANBET | Miss E. Williams |
| YAMSAR | Mrs. A. Ramsay |

All these names contain letters from their breeders' surnames, in some cases completely (BRIGIDEER, BYRNETTS, KATHSHORT), as a full or partial anagram (CHALMI, ELSILRAC, LANDOSARLA, PEACHLYNN, TAMBEMA, YAMSAR), or as a spelling pronunciation of the owner's initials (EMBEE, ESSAYCI). Mr. and Mrs. Drury have their family name virtually intact, albeit in its plural form. Other names clearly refer to first names: ADNIL (reversal), ARCHSUE, BARCHRIS, CLYBET, DELLOUISE, DENDORIS, DORISMUR, GEOFFLEEN, GILLYANNE, JACQAVID, JESSMAY, LEOCRIS, LIZANMYK, NICOLYN, PATAJOHN, SARAMAY, SHEILDUN. Many of these suggest a combination of husband and wife names (Archie and Sue, Denis and Doris, Geoff and Eileen, Jacqueline and David, Liz and Mike, Nick and Lynne, Pat and John). Others are perhaps the breeder's own first names (Della Louise, Doris Muriel, Jessica May, Sarah May).

Some breeders are even more subtle. Mr. and Mrs. Williamson-Bell breed Russian Blues, so for their prefix devised KOLOKOL, the Russian for "bell." This name therefore refers to the breed, as more obviously (and in English!) does ESEMAIS, a reversal of "Siamese." CATAMANDA appears to be a combination of "cat" and "Amanda," while VECTENSIAN may refer to a cat (or breeder) from the Isle of Wight, known to the Romans as *Vectis*.

EIRREM is a reversal of the first word of MERRIE CHRISTMAS, in this case the name of the cat's dam.

"Administrative" prefixes, mentioned above, were introduced by the GCCF in 1982. As their name suggests, they are intended purely for administrative purposes, and are devised by the GCCF itself. They are meaningless as words, but significant as coded designations. For the first year, which administratively in fact ran from October 1, 1982, through December 31, 1983, four administrative prefixes were allocated: ADONELO, ADONESH, ADONEBU and ADONEAM. In these, AD stands for "*ad*ministrative," ONE for "(year) *one*," LO for "*Lo*nghair," SH for "*Sh*orthair," BU for "*Bu*rmese," and AM for "Si*am*ese." The latter are the four classes into which the register is divided. For the whole of 1984, the second year of their application, the four prefixes accordingly changed to ADUELO, ADUESH, ADUEBU and ADUAM, the DUE representing "(year) two" (as in *duet,* for example). For 1985 the prefixes altered to ADREELO, ADREESH, ADREEBU and ADREEAM, with REE representing "(year) th*ree*." In 1986 they became ADQWELO, ADQWESH, ADQWEBU and ADQWEAM, with QWE indicating "(year) four" (and suggesting a word such as *quartet*); for 1987 they were ADIVELO, ADIVESH, ADIVEBU and ADIVEAM, with FIVE denoting "(year) *five*," and so on.

A disadvantage of such prefixes is that they are very similar, especially when computerized, and the Governing Council of the Cat Fancy is consequently planning to reorganize their composition so as to make them more distinctive.

Once a registration has been made, a breeder (for cats usually known as a "fancier") cannot add a prefix to a nonprefix. In fact, a name, once registered, can only very rarely be altered, and then only in circumstances that make such an alteration desirable. For example, a kitten may have been registered under the wrong sex, and given a female name instead of a male. In such a case, a name like BIRDSNEST SALLY can be re-registered as (for example) BIRDSNEST SAM. But where a name is essentially "sexless," such as CATAMANDA CAVALCADE, there will be no need to alter it, since it can apply equally to a male or a female.

A historic listing of registered cats' names is of interest, and may be compared with the professional dogs' names given above. The following are included in Frances Simpson's *Book of the Cat,* published in 1903 (see Bibliography), and thus relate to names dating from the turn of the 20th century. The breed names are as cited in the book. Some, such as Silver Persians, are no longer a recognized breed. Others, such as Orange Persians, are now known by a different name, in this case Red Tabby Longhairs.

## 7. Pedigree and Show Names

*White Persians:* LORD GWYNNE, POWDER PUFF, WHITE FRIAR, MASHER, MISS WHITEY
*Blue Persians:* BEAUTY BOY, WINKS, WOOLOOMOOLOO, MOKO, THE MIGHTY ATOM
*Silver Persians:* DIMITY, THE ABSENT-MINDED BEGGAR, CHINNIE, TWIN, I
*Silver Tabby Persians:* SHROVER, TOPSO, LADY PINK, CLIMAX, ROIALL FLUFFBALL
*Smoke Persians:* PEPPER, TIMKINS, RANJI, SILVER SOOT, BULGER, JUBILEE
*Orange Persians:* PRINCE CHARLIE, QUEEN ELIZABETH, GOLDYLOCKS, MARIGOLD
*Cream or Fawn Persians:* DEVONSHIRE CREAM, FAWN, PIXIE, LORD CREMORNE
*Tortoiseshell Persians:* CURIOSITY, PANSY, WALLFLOWER, SNAPDRAGON, TOPSY
*Tortoiseshell and White Persians:* CHUMLY, SUSAN, PEGGY PRIMROSE
*Brown Tabby Persians:* RAJAH, MATER, PERSIMMON, MAZAWATTEE, RUFFIE
*Manx:* D-TAIL, STUMPS, GOLFSTICKS, MONA
*Siamese:* SI, MEO, SIAM, SUSA, AH CHOO, WALLY PUG, CAMEO, TO-TO, YOLANDA
*Russian Blues:* KOLA, LINGPOPO, MOSCOW, OLGA, YULA, FASHODA
*American:* KING OF THE SILVER, JACK FROST, LADY LOLLIPOP, LADY LOLA
*Maine:* DOT, BABA, RICHELIEU, LEO, MAXINE, TAGS

LORD GWYNNE has a name that suggests Welsh *gwyn,* "white." MASHER's name relates to a former term for a dandy. WOOLOOMOOLOO, presumably adopted for its curiosity value, is now the name of a district of Sydney, Australia. CHINNIE was a *Chin*chilla Persian, as an alternate name for a Silver Persian. The cat named I was the twin of TWIN, and had the full registered name of I, BEAUTY'S DAUGHTER, in turn relating to BEAUTY, her dam. PRINCE CHARLIE and QUEEN ELIZABETH have names that look strangely contemporary. But PRINCE CHARLIE was presumably named because he was "Bonnie," while QUEEN ELIZABETH has a name referring to the red hair of the famous "Virgin Queen." LORD CREMORNE, while suggesting a genuine aristocratic title (that of Viscount *Cremorne*), is meant to hint at *cream,* while CHUMLY's name is almost certainly the familiar pun on the aristocratic surname *Cholmondeley,* traditionally so pronounced. PERSIMMON, in this instance, was named for the Derby winner of 1896. (See his name in Chapter 10, page 126.) MAZAWATTEE was a former well-known brand of tea. D-TAIL has a punning name for a tailless cat, as Manx cats are. MONA was the Roman name for the Isle of Man. SI is short for the cat's breed. The name of WALLY PUG was probably suggested by the popular children's book, *The Wallypug of Why* (1895), a pseudo-oriental fantasy based loosely on *Alice in Wonderland*. The cat FASHODA was born in 1896, and named for the Sudanese town (now Kodok) that was then at the center of a territorial dispute between Britain and France. YULA has a name that is the Russian word for "spinning top" or "fidget." The first three American cats listed were all silverhaired. Of the Maine cats, also known as Maine coon cats (see Appendix I), RICHELIEU has a name that may be intended to evoke the *Éminence Grise,* the "Gray Eminence," otherwise Père Joseph, the monk who was secretary to the famous French cardinal, while MAXINE's name appears to comprise a pun on *Maine* itself, to which the letter *x* has been added. TAGS was a "smoke," and undoubtedly named for his "tags," the black tips of his coat hair, which would have contrasted with his white undercoat.

As with the dogs, many of the names relate to the cat's particular breed or coloring, so that some of the Persians and Siamese have generally exotic or oriental names, while the Russian Blues have Russian names.

# 8

# Horse and Hound Names

Just as "professional" names for dogs are registered by the Kennel Club, and for cats by the Governing Council of the Cat Fancy (see Chapter 7), so the names of racehorses are recorded in Britain by the Jockey Club and published by Weatherbys of Wellingborough every two years as *Regisered Names of Horses*.

The Jockey Club is the supreme authority in control of horse racing and breeding in Britain, and was formed in 1751 to regulate racing at Newmarket (where its headquarters remain today) and support the sport generally. Its equivalent in the United States is the American Jockey Club, founded in 1893.

The Jockey Club's naming regulations are in the main concerned with what names are *not* acceptable, rather than with those that are, so that the following nine types of name are disallowed: (1) names appearing in the current book of *Registered Names of Horses;* (2) names registered since the book was last published; (3) names consisting of more than 18 characters or spaces (with punctuation marks counting as letters); (4) names of well-known persons, unless the person concerned has given permission; (5) names that are similar in spelling or pronunciation to ones already registered; (6) names consisting entirely of a single letter, or ending in one (thus implying that a name may *start* with a single letter, such as B MAJOR); (7) names consisting entirely of numbers, in words or figures; (8) names given for obvious advertising reasons, unless the company or organization concerned has given its written approval and the Jockey Club has given its permission; (9) names which would cause any confusion in the administration of a race. To this it should be added that any "indecent" names would equally be unacceptable, even if disguised in the form of an anagram. On the other hand "meaningless" names are perfectly in order, so long as they do not fall into any of the proscribed categories listed above. On the whole the Jockey Club does its best to accept any name that a horse owner chooses. Even so, to guard against the possibility that a name would be found unacceptable, a horse owner is required to give four proposed names in order of preference when completing the official naming form.

We must all be familiar with racehorse names, even if we lack firsthand experience of the track itself or of its betting activity. The names themselves at first glance appear to range from the meaningful (or at least English) to the meaningless (or foreign). Typical are the following runners in the Derby on June 3, 1992: ALFLORA, ALNASR ALWASHEEK, ASSESSOR, DR. DEVIOUS, GREAT PALM, LOBILIO, MUHTARRAM, NINJA DANCER, PARADISE NAVY, POLLEN COUNT, RAINBOW CORNER, RODRIGO DE TRIANO, SILVER WISP, ST. JOVITE, THOURIOS, TWIST AND TURN, WELL SADDLED, YOUNG FREEMAN, YOUNG SENOR. Yet overall one can detect a general "positiveness." ASSESSOR sounds judiciously efficient, DR. DEVIOUS and TWIST AND TURN seem wily, NINJA DANCER appears lively, RAINBOW CORNER has a colorful air, WELL SADDLED could be a potential winner, YOUNG FREEMAN has youth and strength. THOURIOS must surely represent Greek *thouros,* "leaping," "eager." (For the actual origin of RODRIGO DE TRIANO, one of the favorites, see Chapter 10, page 128. The winner turned out to be DR. DEVIOUS.)

Racehorse names are in fact generally favorable or propitious. After all, the horses are there to *win:* they are there to win the race (proving their power, speed, stamina, elegance, intelligence, and the like) and they are there to win money (implying their superiority, class and value). In other words, they epitomize the three ambitions that most of us seek, if we are honest, to achieve: fame, wealth, and health.

The different desirable attributes embodied in racehorse names may be illustrated by the examples below. They are taken from the first issue (May 1974) of the British racing magazine *Tic-Tac,* but could equally have come from any such listing in almost any English-speaking country at any time in the 20th century, with due allowance for the more recent developments of science and technology.

Names expressing speed or stamina are common, and include AIR POWER, AUTO-SPEED, BLITZ, COUP DE FEU, DEMON PATH, EXPLODING, FLYING HERO, FREEFOOT, GO BABY GO, GO TOO, HURRY ROUND, JUMBO JET, MILE-A-MINUTE, PAVE THE WAY, PROPEL, RAPID RIVER, ROARING WIND, RUNNING FIRE, SPACE SHOT, SPEED COP, STORMER, TOP SPEED, WHOOMPH.

Names expressing general superiority can be denoted in various ways. Many owners favor royal names, such as BEAU SOVEREIGN, BLACK REGENT, CARD KING, COURT SENSATION, CROWN CASE, HEAVENLY PRINCE, IMPERIAL CROWN, KING'S FLIGHT, LE DUC, LUNAR QUEEN, MAJESTY, MODEL PRINCESS, PALACE HOPE, PRINCELY MOUNT, QUEEN'S TREASURE, REINE

## 8. Horse and Hound Names

BEAU (a nice pun), ROYAL DANDY, SEA PRINCE, SUN QUEEN, THREE CROWNS, TUDOR SHOON, WHITE PRINCE. In Britain, at any rate, the association between royalty and racing is a real one, since members of the royal family regularly attend races and the Duke of Edinburgh has himself made a reputation as a noted horseman. (Appropriately, as Prince Philip, he has a name that actually means "lover of horses," from the Greek.)

Military names often express the same concept of superiority or class: BRIGADE MAJOR, CAMP COMMANDER, COLONEL NELSON, GENERAL CUSTER, HANDSOME MAJOR, MAJOR ROLE, MARINE PARADE, MILITARY MEDAL, MUSIC MAJOR, SERGEANT ROSE, THE ADMIRAL, VICTOR'S HUSSAR.

Names incorporating colors are equally favored, especially those of a precious metal (*gold, silver*) or a primary color associated with a national flag (*red, blue*). These two last colors also suggest breeding (red blood, blue blood). "Gold" names include BAR GOLD, FLEUR D'OR, GOLD COAST, GOLDEN BEAUTY, GOLD TIPPED, KING MIDAS, LE COQ D'OR, PENSODORO, SPANISH GOLD. "Silver" names are ARGENTAN, BAR SILVERO, KING SILVER, SILVER STRAND, SILVER TIGER. Other colors are represented by such names as BLACK CYGNET, BLUE ACRE, CIEL ROUGE, CORDON ROUGE, FIRST GREY, GREEN SIGNAL, GREY SHOES, IVORY LADY, LILAC WINE, PINK ROSE, RED ALERT, RED CHINA, RED TRACK, ROSE PINK, SPRING AZURE, VERDANT GREEN, WHITE HOPE, WHITE TO MOVE.

The "gold" and "silver" names here of course suggest money, and names implying wealth and winnings are frequently adopted. On the whole they are "livelier" and wittier than most other names, reflecting the excitement and celebrations that a healthy win can bring! They include BLESS THIS HORSE, COMFORTABLY OFF, DOUBLE CHEQUE, FLOATING PENNY, GO FOR BROKE, HIGH AWARD, HUSH MONEY, JOLLY LUCKY, JUST A CHANCE, MINIGOLD, ONLY A MONKEY (the latter being a British slang term for £500), PENNYCUIK (punning on the name of the Scottish town *Penicuik*), PLUCKY PUNTER (i.e., gambler), POCKET PICKER, SHOT IN THE DARK, SMACKERS (British slang for pounds sterling), STERLING LAD, WINNING HAND, WOT AV I MIST.

Names frequently point to a horse's character, reflecting its desirable loyalty, attractiveness, obedience and the like: ALWAYS FAITHFUL, ALWAYS HAPPY, BE FRIENDLY, BEHAVE TOO, BE TRUE, BRAVE LAD, CLEVER PAL, EXEMPLARY, FAIR TACTICS, GENUINE, GOOD COURAGE, HARDY, MADLY GAY, MAGNANIMOUS, MOST APPEALING, PRIDDY NICE, QUICK THINKING, QUITE SWEET, REALISTIC, SMART SAM, SO VALIANT, STRAIGHT AS A DIE, STUNNING, SWELL FELLOW, TEMPERED, UNBIASED, WINSOME (another pun). Of course, many such names could equally be said to apply to the jockey himself, or to his style of riding.

The fact that a horse is a treasured possession, at least to his owner and trainer, is often reflected in a name. "Possessive" names thus include ADAM'S PET, ALF'S CARINO, APPLE OF MY EYE, CASSY'S PET, DEAR ARTHUR, DEE AND ME, FAIR COUSIN, GEOFF'S CHOICE, JACK'S HOPE, LA MIA RAGAZZA (Italian for "my girl"), MAFILLETTE (the French equivalent), OUR HENRY, RONDO'S BOY, SMOKEY'S GIRL, UNCLE CYRIL, WILLIE MY SON.

The relationship that is expressed in such names is either a genuine family one, referring to the horse's own pedigree, or, by implication, the intimate one that frequently exists between a horse and his owner, trainer or jockey. Such a "family" sentiment is particularly strong for a young horse, just beginning its racing career as a colt or filly. The English Classic Races (One Thousand Guineas, Two Thousand Guineas, Derby, Oaks and St. Leger) are intended precisely for three-year-old runners. For this reason, many names relate to a horse's gender as well as its youth. For colts, too, they can also express the desirable attributes of a good jockey, who is himself popularly regarded as youthful, athletic, and boyish. Names for colts thus often include the word "boy" or "lad," the latter itself further suggesting the stable lads (usually grown men) who look after the horses in their racing stables. Typical names of this type are BARROW BOY, BISCUIT BOY, BUGLE BOY, FUNNY LAD, INVISIBLE LAD, MASTER SCORCHIN, MEDINA BOY, MERRY BOY, RUSTIC LAD, STUPENDOUS BOY, SUNNYBOY, YOUNG CATO, YOUNG ROBERT.

Fillies have similarly affectionate names, and frequently include the words "girl," "lass" or "miss": ABERDEEN LASSIE, BROADWAY GIRL, DANCING GIRL, IRRESISTIBLE MISS, JUBILEE GIRL, LINBURY LASS, MEADOW LADY, MISS BY MILES, PARTY GIRL, RELUCTANT MAID, SAINTLY MISS, TOKEN GIRL. ("Miss" and "maid," not surprisingly, are often exploited for a pun.)

A horse's name may also, of course, reflect a relationship in the owner's family or among his or her friends. The British rider Mary Thomson, who came to the fore in the Badminton Horse Trials in the late 1980s, owns five horses named KING BORIS, KING ARTHUR, KING MAX, KING SAMUEL, and KING CUTHBERT. The first word of each of these names is not simply a royal prefix but the surname of her boyfriend, David *King*.

It is only a small step from an affectionate name to one that more obviously expresses glamor or romance, the romance itself being as much in the actual sport of racing as in the "love bond" formed between a horse and its rider. The relationship between the two, after all, is not only highly emotional but literally physical. By any standards, also, a racehorse at its peak is a fine and handsome animal, a "gorgeous beast." Names implying the glamorous nature of it all may sometimes suffer from a certain twentieth-century gloss or glitz, but the sentiments they state are real enough: AGE OF CONSENT, BLUSHING MAID, BOLD LOVE, CODE OF LOVE, CUPID, FIERY KISS, GENTLE THOUGHTS, GREAT LOVE, HEARTBEAT, HONEY LOVER, INVISIBLE ROMANCE, ITS-A-MATCH, JUVANESCENCE, KISSING, LATE LOVE, LOVELY MATCH, ME TARZAN, MINNAMOUR, NATIVE BRIDE, PERFECT MARRIAGE, READY AND WILLING, SECRET DREAM, SEDUCTIVE, SPRING SECRET, SWEET SLAVERY, TWO FOR JOY.

Racing is not only glamorous, it is *fun* (as implied in the original sense of "sport"). For this reason many owners like to indulge in a pun when giving a name. Some examples have already been instanced. A common device is to distort the standard spelling of a word or name in such a way that the pun is (sometimes embarrassingly) obvious. Typical of such names are AURE-U-LUPI, AVEROF, BOLDINI, DESERT CHAT, FELL SWOOP, GOSPILL HILL, HOPE OF HOLLAND, IRRTUM (more recondite, this, as the German word for "error"), LITTLE BO BLEEP, MAKAROVA (Russian names make good puns; compare AVEROF just instanced), MAN KIND, OFF SCENT, ORLON, ROAN RANGER, SYLVANECTE. Of course, as with all such names, there may have been a punning adoption from an existing name. BOLDINI, for instance, may have taken the name of the character so called in P.C. Wren's bestselling novel *Beau Geste* (1924), and MAKAROVA may have been inspired by one or other of the Russian actresses of this name.

Most common of all, however, and in a sense overriding any individual sense or pun, is the tradition whereby a horse's name reflects that of one or both of its parents. This is a practical way of denoting a horse's particular pedigree. If at the same time a punning allusion can be incorporated in the name, so much the better. The pun in such cases may be by either sense or sound, perhaps both. It also makes the name itself more meaningful. DESERT CHAT, for example, has a name that is not simply a pun on "Desert Rat." His father (sire) was DERRING-DO and his mother (dam) was FIERY KITTEN. He thus has the DE- of his father's name (for the sound link), but the CHAT (French for "cat") to denote the KITTEN in his mother's name (for the sense link). And he in turn has a name that will readily lend itself to further variation for his own offspring.

By way of illustration, here are some race winners of the 1973 season. It will be noticed that sound and sense references are not always both present, but that the new name will always in some way reflect the two existing ones.

| *Horse* | *Sire* | *Dam* |
| --- | --- | --- |
| BALACLAVA BOY | RIGHT BOY | FLORENCE NIGHTINGALE |
| ABOVE SUSPICION | COURT MARTIAL | ABOVE BOARD |
| ARCTIC SLAVE | ARCTIC STAR | ROMAN GALLEY |

| Horse | Sire | Dam |
|---|---|---|
| BOLD HOUR | BOLD RULER | SEVEN THIRTY |
| DISCIPLINARIAN | BOLD RULER | LADY BE GOOD |
| GAME RIGHTS | BIG GAME | JUST RIGHT |
| I SAY | SAYAJIRAO | ISETTA |
| JOLLY JET | JET ACTION | LA JOLIETTE |
| PALL MALL | PALESTINE | MALAPERT |
| WILL GO | WILL SOMERS | GO NOW |
| HIGH WIRE | SKY GIPSY | KABLE |
| DROP-EVEN | EVEN MONEY | LET GO |
| BARMY | SILLY SEASON | BEWILDER |
| CAMOUFLAGE | MARCH PAST | HIDING PLACE |
| SMACKERS | DARLING BOY | KISS KISS |

Many references are obvious here. Florence Nightingale was famous for her work during the Crimean War, which included the major battle of Balaclava. "Ruler" means both "person who rules" and "length of wood used for measuring." The latter implement was long used in schools for inflicting punishment (by a rap on the hand, for example): hence "Disciplinarian." "Smackers" (already mentioned above) here has its other slang sense of "kisses."

Knowledge of a horse's pedigree thus in many instances helps to explain its name. This is particularly true of apparently meaningless names, which usually result from a blend of elements in the names of the parents. In certain instances, moreover, a particular element can be traced back through several generations, preserved from sire to son. Thus BOLD REASONING (born 1968) was the son of BOLDNESIAN (born 1963) and grandson of BOLD RULER (born 1954, in table above), who was himself the son of NASRULLAH (born 1940) and grandson of the famous NEARCO (born 1935). Identical elements are also found in the names of the offspring of a single sire. For example, apart from BOLDNESIAN, BOLD RULER had offspring named BOLD BIDDER, BOLD EXPERIENCE, BOLD HOUR, and BOLD LAD, while the major race winner HYPERION, born in 1930 (see Chapter 10, page 119), was the sire of offspring named HYACINTHUS, HYCILLA, HYLANDER, HYPERICUM, and HYPERIDES. In similar fashion, NORTHERN DANCER, born in 1961 as the grandson of NEARCO and sire of the famous NIJINSKY (see Chapter 10, page 125), had offspring who included ALMA NORTH, BROADWAY DANCER, DANCERS COUNTESS, LAURIES DANCER, NORTHERN BABY, NORTHERN TASTE, NORTHERNETTE, and TRUE NORTH.

Sometimes the references may be more subtle, without a simple repetition of a word or element. The Australian racehorse LE FILOU, himself the son of VATELLOR and FILEUSE, has a French name meaning "the thief," "the crook." This meaning is reflected in the names of several of his offspring, who apart from BIG PHILOU and GAY FILOU included CRACKSMAN, LIGHT FINGERS, PETERMAN, and RED HANDED. (A *peterman* is a safecracker.)

Important races such as the Derby and Grand National often attract people who might otherwise never bet to "have a flutter." On such occasions, hardly knowing one horse from another and blissfully ignorant of "form," backers usually regard the horse's name itself as the best guide to a potential winner. That is, if the name seems lucky in some way, the horse could win or be well placed. An example of the half-serious, half-playful attitude to such a gamble was illustrated by an article in *TV Times* ("Your Derby Winner Could Be in the Stars," May 29, 1980) in which television celebrities were asked how *they* picked a Derby winner. Some favored a professional tip, others preferred the random "pin-sticking" method. For many, however, the name was the thing: "If the horse's name doesn't include the initials of the man I fancy most, I'll look for one that does" (Betty Driver); "A nice, round name will do me. Maybe there will be some association with a name. A jockey called Eric, maybe. In Hong Kong I once scooped a 125-to-1 shot on a horse called Eric's Folly" (Eric Sykes); "I can safely put my money on any horse with a name that suggests Fred Scuttle, Ernie the Milkman, The Chinaman or The Milky Bar Kid" (Benny Hill, naming the various characters he had played).

Sometimes such "name magic" really does produce a winner, to the embarrassment of the bookmakers. An outsider called WINDSOR LAD was running in the 1934 Derby at a time when the Prince of Wales, himself a "Windsor lad," was one of Britain's most popular figures. This circumstance, coupled with the prediction by a noted clairvoyant that a horse whose name began with "W" would win the Derby, resulted in widespread backing for the horse. The odds came down to 15–2, WINDSOR LAD cantered home, and the bookies suffered heavy losses. They came equally unstuck on a later occasion. A horse named AIRBORNE was running in the 1946 Derby. So many housewives had husbands, sons or brothers who had served in the Air Force during World War II that thousands of them backed the horse. It romped home at the unbelievable price of 50–1.

For a fictional account of "name magic" coming up with a winner, see SADOWA in part 8 of Appendix III. For the origins of the names of many famous racehorse winners, see Chapter 10, page 105.

Hunting horses are given names on much the same principles as for racehorses, and the following examples are thus from the *Hunter Stud Book* for 1933:

## The Naming of Animals

| Horse | Sire | Dam |
|---|---|---|
| ARDROSE | ARDAVON | PRIMROSE |
| ATHALASS | APOTHECARY | GREY LASS |
| BARBMARK | BARBICAN | MARKOVER |
| BODANE | BODDAM | SULTANE |
| CYLGAL | CYLETTE | YAFFORD GIRL |
| DOLLETTE | ETON BOY | DOLLY |
| FURETTE | FURORE | FLANNELETTE |
| GAYDIER | BOULEVARDIER | GAYLARCH |
| LIMSY | LIMOSIN | DAISY |
| MINCORA | CORBRIDGE | MINDIBS |
| VODEIRA | BALLYVODOCK | MADEIRA |

The fact that the name of sire or dam (or even both) may also be a random blend makes it rather unlikely that the resulting name will be meaningful. (MINDIBS is such a name, and resulted from the renaming of a mare originally called MISS DIBS.) It must be said, again, that apparently meaningless names may actually have some sense, even if in a language other than English. ATHALASS, for example, has a name which, while ostensibly composed of letters from APOTHECARY and GREY LASS, happens to suggest Greek *athalassos,* "inland" (literally "without sea"). But this may be over-ingenious.

A particular hunt will have not only its horses but its hounds; here too, traditional principles lie behind the names.

The names of hounds are mostly given as for show dogs, that is, their names will be linked by a common theme. The link is often simply by way of an identical first letter. By way of example, here are the six litters of foxhound puppies who appeared at a puppy show in 1974 held in England by the Avon Vale Hunt, with kennels in Wiltshire. The litters are listed in chronological order, so that Litter 1 were whelped in December 1972 and Litter 6 in June 1973:

*Litter 1:* (dogs) GLASGOW, (bitches) GLEAMING, GLEEFUL, GLISTEN
*Litter 2:* (dogs) GIMCRACK, GIMLET, GILDER
*Litter 3:* (dogs) GILBERT, GIRDER, (bitches) GIGGLE, GIRLISH, GIFTED, GIRDLE
*Litter 4:* (dogs) ARCHER, ARGUS, ARSENAL, (bitches) ARDUOUS, ARTICLE, ARTLESS, ARDENT, ARIEL, ARMLET, ARSENIC
*Litter 5:* (dogs) GIMBAL, GIMMICK, GINGER, (bitches) GILLIAN, GIPSY
*Litter 6:* (dogs) CASPIAN, (bitches) CABLE, CAPSULE, CANDOUR, CAMERA

Litter 1 were sired by GLANCER, with PADDLE as their dam. Litters 2 and 3 were both sired by GIMCRACK (from the Duke of Beaufort's hunt), but the dam of Litter 2 was GRAPHIC, while that of Litter 3 was GLOAMING. Litter 4 were sired by ARKLE, with CRISIS as their dam. Litter 5 were again sired by GIMCRACK, with the dam this time GLOOMY. Litter 6 was finally sired by CAPTAIN, while the dam was RAPTURE. As can be seen, the puppies retain the first letter of the name of their sire, not of their dam. The names themselves are a mixture of meaningful (GLEEFUL, GIRLISH, ARDUOUS, GINGER) and conventional or essentially arbitrary (GIMLET, ARTICLE, CABLE, CAMERA). Hound puppies are sometimes named after famous horses, as here can be seen for GIMCRACK and ARKLE, two noted racehorses.

The hounds listed here are only puppies and not actual working hounds as yet. They are therefore subject to being sold and having their names changed, like any other professionally bred dog. In fact the names of working hounds are very similar, and are the usual blend of meaningful and conventional names. No attempt is normally made to associate the names of hounds within a single pack.

The following are the hounds in the Royal Artillery Hunt, England, for the season 1971-72, when the pack consisted of 33 couple (30 doghounds and 36 bitches). The hounds are listed in order of seniority, with the oldest, the bitch TRUELOVE, having been "on active service" for 10 years.

*Doghounds:* TRACEY (*sic*), MOHAWK, PLAYBOY, PLEADER, PRESIDENT, BRIGAND, PLANTER, PSALMIST, PILLAGER, POSTMAN, SAILOR, SAMPSON, SHELDRAKE, SPARTAN, PAGEANT, PILGRIM, PINTO, FLAGMAN, PLOUGHMAN, POACHER, POLESTAR, PROCTOR, PANTHER, PARSON, PATRICK, PLANET, MANAGER, MARKSMAN, MESSMATE, MERRYMAN

*Bitches:* TRUELOVE, FEATHER, PANSY, PATCHWORK, DAINTY, DAIRYMAID, FANCY, PORTRAIT, POSY, PRIMROSE, PROMISE, PRUDENCE, BRACELET, BRACKEN, PANCAKE, PASTRY, PICNIC, MELBA, MERCY, PHANTOM, PRACTICE, PRICELESS, PRODIGAL, FAIREST, FAMOUS, PLOVER, POSITIVE, PROSPECT, PAMELA, PARITY, PASSION, MABEL, MEGAN, MANGLE, MAGPIE, MATCHBOX

Many of the hounds are closely related. For example, all the hounds whose names begin with M are the litter of PLANTER and MELBA, this time using the initial of the dam's name. PLANTER, PSALMIST and PANCAKE were sired by CRICKETER, a hound from the Heythrop Hunt. Earlier, the sire of DAINTY and DAIRYMAID had been SHELDRAKE, of the Beaufort

Hunt. His name was in turn adopted for a doghound of the Royal Artillery Hunt.

TRACEY may seem an unusual name for a male hound. But the dog joined the pack in 1963, at a time when *Tracy* was only just beginning to become popular (and then at first in the United States) as a girl's name. For a hound, the name suggests the dog's ability to *trace* a scent. As can be seen, however, the names of the bitches are markedly more "feminine" than the doghound names, despite the representation of conventional names.

Some of the names of the doghounds (but not the bitches) are "businesslike" in the sense that they denote an occupation or profession. Among such names are PRESIDENT, PLANTER, POSTMAN, SAILOR, FLAGMAN, PLOUGHMAN, PROCTOR, PARSON, MANAGER and MARKSMAN. But the occupations are almost all arbitrary, although PLANTER or MARKSMAN could relate to the hunting work of a hound. So could FLAGMAN, if the name is a punning reference to the dog's "flag" or tail.

The Avon Vale puppies listed above happened to have names based on that of their sire. Where an identity of letter is observed, however, the name may relate to either sire or dam, sometimes switching fairly randomly from one to the other down the generations. Thus the hound DAFFODIL, born in 1939 and in the Duke of Beaufort's Hunt, had MUSTARD (born 1932) as her sire and DEWDROP (1933) as her dam, while DEWDROP had WARRIOR (1929) as her sire and DIMPLE (1928) as her dam. DIMPLE, however, had DREAMER (1925) as her sire and PAMELA (1922) as her dam, and DREAMER's parents both had names beginning with D, respectively DANGEROUS (1920) and DINAH (1922). DAFFODIL herself mated with SARACEN (born 1939) to whelp SALESMAN in 1944.

Not all hunts rely simply on the identity of letter, however. Some packs name litters after flowers and birds, for example, while earlier this century the Blackmoor Vale Hunt, Somerset, had hounds with names of pure pastoral poetry, such as DOVECOTE, GREENWOOD, LILYWOOD, WAKEWOOD, WILDNIGHT, WILDROSE, and WILLOW. By contrast, another hunt, which must remain nameless, had a pack of foxhounds that included HOOLIGAN, HORRIBLE, HORROR, and HOVEL.

It will be noticed that almost all these names are of two syllables, with the accent on the first syllable. The practice of giving hounds names of this type is traditional, and is designed for practical reasons. A name of this kind is easiest to pronounce and is (theoretically) easy for an individual hound to distinguish. Names of more than two syllables with the accent on the second syllable, such as "Outsider" or "Veronica," are thus unsuitable. However, masters of some hunts give or allow names that may obey these principles but be unsatisfactory in other ways. In some cases dogs are given very similar names, which are difficult to distinguish when spoken. Examples are the dogs GIRDER and GIRDLE in the Avon Vale Hunt (above). In others, names are given that are easy to pronounce or distinguish, but that are unsuitable or even offensive in meaning. The four names quoted at the end of the previous paragraph are hardly complimentary.

The bestowal of inappropriate names is nothing new. Charles Lennox, 3rd Duke of Richmond (1735–1806), had a pack of hounds that included MADNESS, MILDEW, MISERY, and MURDER, while the famous English sportsman George ("Squire") Osbaldeston (1787–1866) had hounds called BOOZER and HERNIA. Some hunts went in for esoteric, even eccentric names in the mid–19th century. Among those recorded for the famous Berkeley Hunt, Gloucestershire, at this time are COLOCYNTH, HIPPOGRIFF, HOWITZER (this in 1847), and WOWSKI. Misnamings have also occurred. The celebrated master of foxhounds Lord Henry Bentinck (1804–1870) had an outstanding hound, described by himself as "probably the best dog that ever ran in the Midland Counties," called TOMBOY. Yet this name properly belongs to a female hound, not a male! (The hounds belonging to Lord Bentinck are described in detail in Lord Charles Cavendish-Bentinck's book *Lord Henry Bentinck's Foxhounds,* published in 1930.) Sometimes the hound himself discredits his name. The story is told of a bitch hound named DECENT who belonged to master of foxhounds Sir Edward Curre (1855–1930). She was tempted to "tread the primrose path" and was seen to be obviously in whelp (pregnant). Sir Edward called her to him and was heard to mutter: "Decent! Decent? What is the use of me trying to give you dear little things nice names?"

The eminent English sportsman Peter Beckford (1740–1811) was the first master of foxhounds to describe the whole system of hunting in writing, and his *Essays on Hunting* (1781) contain a wealth of anecdotes relating to the names of hounds. After telling of such inapposite names as GIPSY for a white hound, RUBY for a gray hound, SNOWBALL for a dark-colored dog, and BLUEMAN for a hound that was "of any colour but blue," he recounts the tale of a hound named LYMAN. "Lyman!," exclaimed the master of hounds, on hearing his name, "Why, James, what does Lyman mean?" "Lord, Sir," replied James, "what does *anything* mean?" Another story concerns a hound named MADAM. The dog consistently ignored the huntsman's reprimands, whereupon she was vigorously scolded with the words "Madam, you d----d bitch!" This recalls a similar episode in the life of "Otter" Davis, curate and whipper-in to the celebrated "sporting parson" Jack Russell (see his name in Appendix I). An angry old woman took Davis to court for repeated insults when, on an early morning otter hunt near her cottage by the Avon River, he encouraged his favorite hound MIDNIGHT with such exhortations as "Get up, you old bitch!" and "Yooi over, old girl!"

Despite the eccentricities noted here, many hound names have remained in regular use for several centuries now. Among those listed by Nicholas Cox in *The Gentleman's Recreation* (1674) are BLUEMAN, BONNY, DAMOSEL, JEWEL, JUNO, LADY, MAULKIN, MERRY BOY, MOTLEY, MUSICK, SINGWELL, TOUCHSTONE, and TRUELIPS, while a contemporary publication, *A Catalogue of some general Names of Hounds and Beagles,* has SOUNDWELL, TATTLER, THUNDER, and TICKLER. Of course, topical names are also given as and when the occasion arises. Stud books show that WATERLOO and WELLINGTON were fashionable in their day (1815), while BATTLESHIP, CHAMBERLAIN, CHURCHILL, CONVOY, FRANCO, and SPITFIRE appear in the years of World War II, with D-DAY found in 1945, VICTORY common in 1946, and SPUTNIK and TEE-VEE recorded in the late 1950s. (The use of WATERLOO as a hound name was possible because the placename was stressed on its first syllable at one time always.)

The matter of foxhound nomenclature is considered by "Yoi-Over," the pen name of a retired English kennelmaster, in *Hold Hard! Hounds, Please!* (1924). Emphasizing the desirability of giving short, distinct names to hounds, the writer indicates the importance of selecting a name that is not readily mispronounced, especially by a huntsman or kennelman who may normally speak a regional form of English. He mentions cases where one bitch named FAVOURITE had her name pronounced by the kennelman without the final *t,* and where another called HEROINE was regularly addressed by the kennelman as "Irwin." On the other hand, admits "Yoi-Over," a dropped *h* is less significant, since a hound named HANDOVER will respond readily enough to "Andover." The author's aversion to lengthy hound names obviously sprang from personal experience, since he tells how he regularly had to summon hounds named MONADELPHIA and HELLEBORIUM "under the stress of munching my crust of bread and cheese on the way home to kennel." "Mona" and "Hetty" would have been far easier, he comments! (The names themselves are apparently of classical botanical inspiration.) On a somewhat different tack, "Yoi-Over" tells of a huntsman who would change a hound's name if he did not like it. On one occasion, he altered the name PLATO to PLAYTO, thinking his master had made a spelling slip. On another, the same man was instructed to take a terrier named KING FROST with the hounds to a show. He took two terriers, however, and at the show, over their respective heads on the bench, duly wrote KING for one and FROST for the other.

For a full listing of traditional hound names, see Appendix II.

Among the best known working horses in Britain are the drays employed by Whitbread's Brewery in London. They still regularly leave their stables in Chiswell Street to travel over a five-mile radius while serving as a mobile promotion for the brewery's beer. The horses are always harnessed in pairs and named in pairs. In the 1970s, the pairs comprised the following couples: SAILOR and TRIDENT, TIME and TIDE, ROYAL and SOVEREIGN, UNION and ULSTER, WALES and WHISSENDINE, WINSTON and WASHINGTON, MIGHT and MAIN, THUNDER and LIGHTNING, PRIDE and PREJUDICE, QUAKER and ROBERT, POMP and CIRCUMSTANCE, CROWN and ANCHOR, XERXES and ARTAXERXES.

Most of these pairs have obvious associations, although WALES and WHISSENDINE seem to have a link of alliteration only, as do WINSTON and WASHINGTON. (Whissendine is a village in Leicestershire.) QUAKER and ROBERT may have some religious link: Robert Barclay was a noted Scottish Quaker. XERXES and ARTAXERXES, father and son, were famous 5th century BC kings of Persia. More to the point here, however, they are also the names of two horses in the comic novels of the sporting writer Robert Surtees (see Appendix III). POMP and CIRCUMSTANCE have the name of the march by the composer Edward Elgar that famously gave the patriotic song "Land of Hope and Glory."

By the mid-1980s the Whitbread drays were down from 14 pairs to 6. Three of the pairs were the same: TIME and TIDE, PRIDE and PREJUDICE, POMP and CIRCUMSTANCE. The other three were: HIGH and MIGHTY, PIKEMAN and MUSKETEER, AJAX and ACHILLES. These all have familiar associations, with AJAX and ACHILLES being not merely mythological figures but heroes who fought on the side of the Greeks in the Trojan War.

Names like these are simply randomly artistic, and are not even generally descriptive.

Police horses are found as working horses in many cities of the world. Their names are often random or "regimented," with little significance for an individual animal. In London's City of London police force, for example, horses are named after districts of Inner London, such as BARBICAN and WALBROOK. By contrast, horses working for the Metropolitan Police Force (i.e., in Greater London, but not the City of London) are named for the year in which they are acquired, the particular year being indicated by a letter of the alphabet. The horses are bought in batches twice a year, in spring and fall, with roughly 20 horses in each batch. Letter N, for instance, was the one for 1985, when the 21 horses in the first batch were allocated the following names: NAUTIC, NEGUS, NELLIE, NELSON, NEWBURGH, NEWMAN, NEWSBOY, NICHOLA, NICHOLAS, NIJMEGEN, NIMBUS, NIMROD, NINA, NOBLE, NORFOLK, NORMANDY, NORSEMAN, NORTH STAR,

NUBIAN, NUMERAL, and NUTCRACKER. Many of these names are arbitrary, but some have a specific reference. Thus NEWBURGH was bought from the place of this name in North Yorkshire, while NEWMAN was named for the then Commissioner of Police, Sir Kenneth *Newman*. A horse bought three years earlier had similarly been named KENNETH for him.

Occasionally a police horse retains the name he already had when acquired. An example is SAFETY BLADE, who was a racehorse presented as a gift. Another is CASAMAYOR, also a former racehorse. But these are the exception to the rule, and the names are almost always given within the Metropolitan Police, with the Assistant Commissioner usually doing the actual naming.

Names that are more generally meaningful are those of police dogs. In a run of some 900 names of dogs that were or had been in the service of the Metropolitan Police Force in 1985, many are arbitrary. On the other hand, some are aggressively significant, as might be expected, and include "power names" such as ACHILLES, ATTILA, BLITZ, BRUISER, CEASAR (*sic*), COPPER, FLAME, FURY, GUNNER, HERCULES, IRONSIDE, JUGGERNAUT, MASHER, PANZER, ROMMEL, SABRE, SOLDIER, STRIKE, THUNDER, TOMAHAWK, TROJAN, WARSPITE and WOLF. The name of CEASAR, however spelled, suggests not only the Roman emperor but a dog that is a "seizer," while COPPER has a name eminently suitable for a dog working with a police force, as well as himself being one that "cops" or catches. One dog, not listed here, is called WEEDY, but this may be either a sarcastic name or an "incident" name, for a dog that at first appeared thin or cowardly.

Certain stock names have proved popular for police dogs over the years, and the following is a record of the "Met" favorites, with the number of dogs who have borne the particular name more than 10 times:

REX (217), PRINCE (215), KIM (106), BRUCE, MAX (88), MAJOR (84), SABRE (73), REBEL (68), SHANE (67), BEN (65), SAM (56), JASON (50), RINTY (43), FLASH (35), SANDY (33), SIMBA (32), ROCKY (28), RICKY (25), KING (24), RUSTY (23), LADDIE (21), RAJAH (20), CEASAR (17), BOB, PADDY (15), BRANDY, PETER (14), LUCKY (13), KAHN (*sic*), SHEP (12), CHARLIE, JET, KELLY, MARCUS (11).

Police dogs are mostly German shepherds, renowned for their speed, stamina and strength. Many of the names imply a general superiority (REX, PRINCE, MAX, MAJOR, KING, RAJAH), while others relate to speed (FLASH, JET), strength (SABRE, ROCKY), or color (SANDY, RUSTY, BRANDY). It is clear that some of the names, including a few of those just quoted, have been adopted from specific dogs. RINTY, for example, is the short name of the famous German shepherd RIN TIN TIN (see Chapter 10, page 128), who became a movie star. SIMBA has a name representing Swahili *simba,* "lion." SHEP, normally associated with sheepdogs, here relates to the breed, German *shep*herd. The relatively low scoring for KING (compared to the top-scoring REX) may result from the fact that the role of this otherwise popular name has been taken over by KIM. If so, this would give "royal" names the top three places. Rather surprisingly, however, the original listing contains not even one DUKE.

# 9

# Farm Animals

Farm animals represent a sort of halfway house between the "professional" animal and the household pet. Farms formerly had (and a few still do have) working animals, mainly in the form of draft horses. Today most farm animals "work" by providing food and clothing: cows for milk, cream, cheese and beef, calves for veal, sheep for wool and mutton, lambs for wool and lamb, pigs for pork, ham and bacon, hens for eggs and meat, turkeys for meat, and other animals such as goats, geese and ducks for a combination of these products. Individual animals of these species frequently become pets. True farm pets are mostly the ubiquitous dogs and cats, with dogs, however, often also working to drive sheep or cattle, and cats coopted as mousers.

Here we are primarily concerned not so much with the pets as with the multiple animals, kept in herds and flocks. How are they named?

Today many large herds of cattle or flocks of sheep contain animals that are increasingly being identified individually by numbers rather than names. Where names are used, they are mostly either in the form of group names, as discussed in Chapter 6, or as traditional names for the particular kind of animal.

Many farm animals are named for their coloring or markings. One Yorkshire cattle herd, with body markings resembling insects, birds or islands on a map, had names such as FLY, GANNET, HERON, SCILLY and LUNDY. Later additions to the herd were also given "map" names, such as AUSTRALIA, ARABIA and ITALY. The resemblance to natural objects was broken by NUFFIE, however, a cow with a body marking that to the farmer seemed to be the outline of three cylinders, suggesting his *Nuffield* tractor. JIG was similar, having a marking suggesting the rounded "lock" on a jigsaw puzzle.

Sometimes names of a herd are a mixture of the generally descriptive and the traditional. A Gloucestershire cattle herd thus had names such as SPOT, SMOKY, DARKIE, BUNNY, CURLEY (*sic*), PANDA, TOPSY, TINY,

WHITE, TWINKLE, PATCH, DOLLY, DAINTY, MAGPIE, REDHEAD, ROSIE and BLUEY, this last being a local word for a black cow with white patches.

Herds and flocks are also frequently given names already used for animals in a previous group. This same Gloucestershire herd thus had cows named KITTY, NOREEN, RINGLET, PIXIE, JEWEL, SAMANTHA, PRINCESS and JULIET, earlier borne by other cows.

Herd names beginning with an identical letter are also popular. In such cases the letter itself is often that of the farmer's own surname or first name, or the initial of a member of his family. One Devonshire farmer named all his South Devon cows with an initial R, for his own first name, *R*ussell, all his Ayrshires with a J, for his wife's name, *J*udith, and all his Friesians with an N, for his daughter, *N*aomi. His threefold herd of 76 cows thus had names as follows:

*South Devon:* RADEGUND, RAE, RAGUSA, RAINBOW, RAINE, RAPUNZEL, RAQUELA, REBELLA, REGAN, REGINA, REMBOLA, RENUNCULA (*sic*), REZEN, RHEBA, RHODA, RICA, RICHENDA, ROBERTA, RODERICA, ROLLO, ROMA, ROMANCE, ROSALIND, ROSE, ROXANNE, ROYALE, ROYDA, RUE;
*Ayrshire:* JABBERWOCKY, JACARANDA, JACKIE, JACOBA, JACQUELINE, JADE, JAEL, JANE, JANET, JANICE, JANINE, JAQUETTA, JASMINE, JAY, JEHENNA (*sic*), JEMIMA, JEMMA, JESCA, JESSE, JESSICA, JESSOP, JET, JEWEL, JILLIAN, JOAN, JOANNA, JOCELYN, JOCUNDA, JOLA, JONQUIL, JOSEPHA, JOY, JUANITA, JUDAH, JULIE, JULIET, JUMAN, JUNIPER, JUNO, JUSTINA;
*Friesian:* NADIA, NANA, NANCY, NANNETTE, NELLIE, NICOLA, NINA, NONA.

The selection of a single letter of the alphabet for naming purposes on farms can be taken to almost obsessive length. A Sussex farming family named Powe who lived at Pythingdean near Pulborough not only named all the cows with an initial P but used this letter for their bull (PERCY), their boar (PUNCH), their Labradors (PEDRO and PANCHO), their poodle (POO, from his breed), their parrot (PHREAD), and even their cottages (PRIMROSE, PENELOPE, PRYM and PINETON).

Some herd names appear arbitrary at first sight but turn out to have a particular theme or origin. The following are the names of some of the cattle on a farm in Kent: AMBER, APRICOT, ASTER, BALTIC, BOUQUET, BRANDY SNAP, CACTUS, CHARCOAL, CIRCUS, EMERALD, FUDGE, KESTREL, LINNET, MADAM, MARIGOLD, MODEL, MUSHROOM, PEACH, PEWTER, POMEGRANATE, ROMANY, SABLE, SAFFRON, SWAN. They seem to represent a random selection of colors, flowers and similar names, but were actually selected by the farmer's wife and daughter who went to the local store and

## 9. Farm Animals

noted the brand names of stockings on the stocking counter, some 60 names in all. As commercial names for a female item of clothing they are therefore eminently suitable for a female animal.

There is no doubt, however, that traditional girls' names and flower names are popular for cows, with the occasional purely descriptive name thrown in. A Somerset farm, for example, had cows named BLUEBELL, BUTTERCUP, CALF-CHASER, DAISY, GLADYS, HOPPY, ROSE, SABRINA and STRAWBERRY, while cows on a Norfolk farm were APRIL, BLUEBELL, BRAMBLE, BUTTERCUP, CANDY, CLOVER, DAISY, EMMA, MOLLY, PANSY, PEGGY, POLLY, ROSE, RUBY and SNOWDROP. A farm in north Staffordshire had cows as follows: BERTHA, BLUEBELL, BUTTERCUP, CAROLE, DAFFODIL, DAISY, DINAH, DOLLY, DORIS, HAZEL, HEATHER, IRIS, JANE, JENNY, JILL, JUDY, MARY, PANSY, PENNY, PHYLLIS, POLLY, PRIMROSE, ROSY, RUBY, SUSIE and VIOLET. In the south of the same county, cows on a farm were named thus: ANNIE, BLUEBELL, BROWNIE, BUTTERCUP, CANDY, DAISY, DEBBIE, DORA, EMMA, FAY, FLORA, FOXGLOVE, GAY, HEATHER, HESTER, HILDA, HONEY, HUMBUG, JANE, JANET, JULIA, KATE, KIM, LAURA, LIZZIE, LUCY, LULU, MAISIE, MARIE, MATILDA, MEG, MIDNIGHT, MINT, MOIRA, MULBERRY, NANCY, NELLY, NETTLE, PATSY, PEARL, POPPY, ROSY, RUTH, SALLY, SUE, TWINKLE, VIV and WENDY. Such names are almost always conventional, although in the latter herd VIV was named for the cowman's girlfriend.

Some herds include names that are really descriptive nicknames. The following are the cows on a farm in west Wales: APRIL, AUNT NORAH, BERYL, BIG BERTHA, BUTTERCUP, DUFFER, ETHEL, FREAKY FANNY, FUNNY FACE, GERALDINE, GLADYS, JOSIE, LITTLE DIAMOND, MABEL, MILLIE, MOLLY, PARANOID PAM, PEARL, RUBY, SARAH, SUNSHINE, TESSA, WATTY, ZULU. Calves on this same farm include: AGNES, BOBBY, CHOCOLATE, DAISY, KALI, KARMA, KIRSTY, KRISHNA. The Hindu names here presumably refer to the fact that, to the Hindus, the cow is a sacred animal. (But did they escape sacrifice on this farm?)

Listings of these and similar cow herds show that names such as BLUEBELL, BUTTERCUP, CLOVER and DAISY turn up again and again, if only because these are meadow and woodland flowers, where cows graze. A frequent name for a bull, whether resident on a farm or an annual visitor for purposes of "professional services rendered," is FERDINAND, presumably from the one in the famous children's picture book *The Story of Ferdinand* (1936) by the American writer Munro Leaf. (For more about him, see Chapter 12, page 148.) However, other names are also popular, and a Scottish farmer reports that the bull who comes every year is always called CHARLIE.

Proof that names are recycled and repeated for farm animals is given by a farming family near Bristol, in the West of England, who say that they have had 16 cows named AUDREY, 12 named FORTUNE, 11 called LILY, and 10 named ASTER.

Names for other groups of farm animals tend to be rather more original, perhaps to show that they are not part of the "common herd" (i.e., of cattle). The following are sheep on a Surrey farm: HAZELNUT, HENRY, NIMMO, PERCY, SINGLETON, SOLO, TOPSY, VELVET, WILLY-BOY, ZEBEDEE. The scope is wider, of course, since both sexes are involved. On this farm, VELVET was a ewe, who lambed twins as follows: SILKY and SATIN, WOOLLY and TWEED, FLAX and HESSIAN, GEORGETTE and NYLON. Cattle on the same farm were named more or less conventionally, however. Pet lambs reared on a Gloucestershire farm included BILL, BUSTER, LUCY, PAM, SARA, SNATCH, SNITCH, STIFFIE (presumably from his gait), and TOBY. They were among a total of 33 bred from a ewe named GRANNY (but originally JIM).

The following are the 20 pigs on a Dorset farm: ARTFUL, BETTY, BUTTERCUP, DAISY, DEBORAH, JUDY, LASSIE, LIGHT-O-DAY, MYRTLE, PIGGIE, PORKIE, PRIMROSE, PUNCH, QUEEN, SNOWBALL, SPOTS, SURPRISE, TOBY, WIGGINS, WONDER. Perhaps LIGHT-O-DAY was so named because he (or she) almost never saw it at birth. Pigs on a farm in north Wales, on the other hand, were named more systematically with the same initial letter: LANTERN, LAURA, LAVINIA, LEAH, LILAC, LINDA, LOLITA, LORNA, LOUISA, LUCY, LUSTRE, LYDIA. The boar on this farm was named HAMLET, as a genuine "pedigree Danish" (to say nothing of his meat potential).

A Gloucestershire farmer recorded the names commonly found for workhorses in the West of England. They included the following:

*Mares:* BEAUTY, BONNY, BOUNCE, DARBY, DARLING, DIAMOND, FLOWER, FROLIC, JENNY, JET, KIT, LIVELY, POPPET, SMILER, TRIMMER, VENTURE, VIOLET, WHITEFOOT;
*Geldings:* BLACKBIRD, BOXER, CAPTAIN, COLONEL, DARKIE, DICK, DRAGON, DRUMMER, DUKE, DUMPLING, GILBERT, JOLLY, MILLMAN, PRINCE, PUNCH, SAM, SHORT, TURPIN.

These are mainly descriptive names, whether directly (BONNY, LIVELY, MILLMAN, WHITEFOOT) or allusively (BOXER, DARBY, DUKE). PUNCH is a name for a stout or squat horse (as in the breed *Suffolk punch,* see Appendix I). DICK was presumably associated in some way with TURPIN, to produce the name of the famous (or notorious) English highwayman. JET would refer to color, not speed.

## 9. Farm Animals

Motives for the naming of individual farm animals, or of small groups of animals, are many and varied. As for the domestic pets in earlier chapters, let us take the individual names first. The quoted words are the owner's explanation for the name(s).

A.J., a cow purchased from a Mr. A. Jones
ALICE, a house-reared lamb; owner is Lewis Carroll fan
ALPHA, a calf, the first to be successfully delivered by the farmer's son
ANOREXIA, a pig who would not take food
ARIEL, a cow, the daughter of DAZ (both brands of washing powder)
AUL' MITHER, the oldest ewe in a Scottish flock
BELL, a very thin hen; a shortening of BELSEN
BERYL THE PERIL, a pig, who was always breaking through the electric fence
BETSY, a hen with a ruff, like that worn by Queen *Elizabeth* I
BOO, a lamb, who became timid when frightened by children
BOREDOM, a Welsh *boar,* with "only four sows to keep him happy"
CAPTAIN, a workhorse
CECIL, a pet lamb; for a bad-tempered farmhand
CHANTICLEER, a bantam cockerel, named by its young owner (whose mother comments, "We have romantic children")
CHARLES, a lamb, the 40th to be born; for the restaurateur Charles *Forte*
CHARLIE, a boar bought with an overdraft; for the owner's bank manager
CRUMPET, a bantam cockerel, who "thinks of nothing else" (i.e., a mate)
HYENA, a cow who as a calf paced back and forth in her pen
JIM MOO, a red cow who supplied the household with milk
LUCKY, a cow, the 13th to calve
MAGIC, a cow, who when being milked always managed to shake her chain off
MARINER, a calf born with an old-looking face (like the Ancient Mariner)
MR. CRAVEN, a ewe; for a local veterinarian
MR. PECKY, a cockerel
MOSTLY, a farm pony; in full, MISS MOSTLY BROWN, for her color
NANCY, a bullock who disdained the opposite sex
NEW FOREST, a cow who had suffered New Forest disease (an eye infection)
NINA, a lamb, the *nin*th to be born
OLD FRED, a gentle-natured boar
OLIVER, a gander, who "attacked from the rear" like Oliver Cromwell
OTHELLO, a black lamb; for the black Shakespearean character
PHYL-A-PAIL, a cow who is a good milker; her daughters are similarly PULL-A-PINT and DRINK-A-PINT; latter name recalls British advertising slogan, "Drinka pinta milka day"

PIGOLETTE, a hand-reared piglet
PITTARD, a pig, for a child's imaginary friend
POPPY, a Jersey cow, who on arrival ate large amounts of poppies
PUSPOTS, a cow who was a summer mastitis victim
RAT, a vicious cow
RUSTY, a cow, named not for color but for her "rusty moo"
SANCTITY, a heifer having a five-teated udder (pun on French *cinq tétines*)
SKINHEAD, a cow, a "juvenile delinquent"
SLASHER, a cow who had gashed open the knee of the farmer's wife
SQUEEKER (*sic*), a pet pig
STRANGER IN PARADISE, a sheep found abandoned on the road; for the popular song
SUSAN SOW, a sow, the favorite of the farmer's daughter, *Susan*
SUTTON BONINGTON, a cow; for her place of purchase (in Nottinghamshire)
TALBOT FLICKER, a Labrador working dog; first word for the family crest, which bears the head of a *Talbot* dog; second word to match name of another dog, himself named for *Flashman*, a character in the school story *Tom Brown's Schooldays,* popular with the owner's children
TEDDY, a "cuddly" cow
THANKSGIVING, a calf born after an epidemic of foot and mouth disease
TOBY TUP, a South Downs tup (i.e., ram)
TOM JONES, a lamb, the 76th to be born; for the popular singer, whose name rhymes with the popular song title, "Seventy-six trombones"
TUT, a cow resembling an ancient Egyptian painting (as for *Tut*ankhamen)
WOODY, a cow, "thick as two short planks"
ZAC, a gander, born on An*zac* Day (April 25, a public holiday in Australia)

Some owners, or at any rate namers, apparently lack inspiration when it comes to giving a name. One farming family named their gray mare MRS. X because they could not think of a name at all. Another family called their black mare merely B.M. In a third family, a cow who had been weak as a calf, and unable to stand without human assistance, was known by her owner simply as THE ONE I LIFTED. At least the skewbald mare CORAL had something like a proper name, although her owner confessed that "the name just suited her," so that it was barely meaningful.

Fortunately, most namers are more imaginative than this.

A sort of compromise between a single name and a group name is a collective name for a number of animals. A family who owned some piglets thus called them THE DAWN CHORUS, without any individual names, while a set of battery hens were known on another farm as THE BALDIES.

The following names have been recorded for pairs of lambs on different farms: BUBBLE and SQUEAK, DAVID and JONATHAN (for the biblical friends), ERIC and ERNIE (for the television comedians *Eric* Morecambe and *Ernie* Wise), ESAU and JACOB (for the biblical brothers), HOLLY and IVY, JACK and JILL, MAN FRIDAY and ROBINSON CRUSOE, MARTHA and MARY (for the biblical sisters), MORECOMBE (*sic*) and WISE (compare ERIC and ERNIE), MUSTARD and CRESS, NUTS and MAY ("nuts *in* May," of course, in the nursery rhyme), PENNY and TWOPENCE, PUNCH and JUDY, SALT and PEPPER, TOM and JERRY, TWINKLE and STAR, WHISKY and SODA.

A pair of Muscovy ducks were named NELSON and EDDIE (*sic*), for the American actor and singer *Nelson Eddy*. Two calves who should have been registered as pedigree stock were duly named ABLE LUNDA and MISS STAEK. Two cows who "always got in a right ol' mess," according to the farmer, when tethered in their stall for the winter, were appropriately (and bluntly) called MUCKHEAP and DUNGHEAP. Two calves were known as BONNY (*sic*) and CLYDE, for the two American desperadoes. A pair of sheep went by the names of BOOTSIE and SNUDGE, for the two army characters, played by Bill Fraser and Alfie Bass, in the 1960s British television comedy series of the same name.

For slightly larger groups of three or more, the following names were recorded: JAMES, JEMIMA and JASMINE, three Muscovy ducks; DOLLY, MOLLY and POLLY, three calves; DOUBLET, MAVERICK and PSALM, three cows (named for well-known racehorses); JAMES BOND, DIVIDEND BOND and PREMIUM BOND, three boars bought from brothers named *Bond;* MAISIE, MARGARET, MARION, MARJORIE and MEGAN, five sheep bought from a man whose name began with the letter M.

On occasions, the origin of the name is unknown or uncertain, even in the family that gave it. The wife of the owner of the ferret GURKASSES comments that the name is one "which I personally find extremely baffling." Since animals' names are spoken much more frequently than they are written, it is highly likely that her rendering of the name is only approximate, anyway. Possibly the name relates to the *Gurkhas,* the fighting race of Nepal. Ferrets, after all, are famous for their fierceness and fighting instinct.

# 10

# *Famous Real Animals*

Many animal names become well-known either because the animals are famous in their own right or because their owners are. Sometimes, however, the two blend, and an animal is famed not only for its illustrious (or notorious) owner but because it is itself widely known. An example of the latter might be CHAMPION, the "wonder horse" of the movies, who became as famous as his actor owner, Gene Autry.

This present chapter is devoted to a selection of such familiar names, whether of racehorses, zoo animals, sporting contestants, movie performers, record breakers, or pets of famous owners. Where the origin of the name is known, it is given. In some cases, however, the name is merely traditional or arbitrary. Even so, its very familiarity will have served to add it to the stockpile of animal names that may be drawn on for future use.

A.P. INDY is the horse that American horsemen hoped would win the 1992 Kentucky Derby but that in the event lacked the necessary literal horsepower. He was owned by a Japanese businessman who loved fast cars as well as fast horses, and his name was therefore aptly "speedy." It derives from *Autopolis*, a 70,000-seat automobile racetrack, and the famous *Indy* 500, the world's most spectacular auto race, held annually on Memorial Day (May 30) at the Indianapolis Motor Speedway.

ABLE and BAKER, two monkeys, were the first animals to be sent into space by the United States (1959). Their names represent the then standard signalling designations of the first two letters of the alphabet.

ABU AL-ABBAS was an elephant presented to the emperor Charlemagne in A.D. 802 by the Caliph of Baghdad. He was kept at Aix-la-Chapelle (Aachen), Charlemagne's second capital, and went with him on an annual tour of the empire. His Muslim name means "Son of the Lion," referring to Abbas, uncle of the prophet Mohammed (Muhammad).

AGRIPPINA was the cat owned by the American writer Agnes Repplier

(1855–1950). Her name looks suspiciously like a blend of *grip* and a near-anagram of her mistress's name.

APOLLINARIS, BEELZEBUB, BLATHERSKITE and ZOROASTER were four cats owned by the American novelist and humorist Mark Twain. In a letter to a children's magazine, the writer explained that he gave the particular names "not in an unfriendly spirit, but merely to practice the children in large and difficult styles of pronunciation." *Apollinaris* was actually a 4th-century Syrian teacher and theologian, *Beelzebub* is a biblical name for the devil, *Blatherskite* is an American colloquial word for a talkative person, and *Zoroaster* was a famous Persian prophet of the 5th century B.C.

ARKLE, the famous Irish steeplechaser, was foaled in 1957. His name partly reflects that of his sire, ARCHIVE. He was more deliberately named, however, for *Arkle,* a Scottish mountain ridge. The name is appropriate for a horse that needs to "aim high" as he clears the jumps. For a similar name, compare FOINAVON (below).

ATOSSA was the Persian cat owned by the English writer and critic Matthew Arnold (1822–1888), who recalled him in his poem *Poor Mathias:* "Rover, with the good brown head,/Great Atossa, they are dead." (MATHIAS was his canary, also of the past, and possibly named for himself.) The cat, familiarly called TOSS, was named for the mother of Xerxes, the Persian king who vainly attempted to subdue Greece in the 5th century B.C. The historic Atossa was thus a Persian queen, as also, in the feline sense, was Arnold's pet, *queen* being a regular term for an adult female cat.

BANDOOLA was the Burmese elephant who gained fame for leading a party of people and animals to safety from Burma in 1944 after the Japanese invasion. His exploits were described in the book *Bandoola* (1955) by Lt. Col. John H. Williams. The elephant was named for Maha *Bandula,* the Burmese general who was killed in the British capture of Danubyu, Burma, in 1825.

BARRE-DE-ROUILLE was the big, orange-striped alley cat owned by the French novelist Joris Karl Huysmans (1848–1907). His name means "rusty stripe," for his color. On his death, the writer acquired another cat, thin, ugly, and gray-striped, whom he called MOUCHE ("fly"). He depicted him in his autobiographical novel *Là-Bas* ("Down There") (1891).

BARRY was the famous rescue dog who was trained in the early 19th century at the hospice in the Swiss Alps near the Great St. Bernard Pass. His name refers to his breed of *barihund,* literally "bear hound," for the dogs' bearish appearance. The breed was officially renamed *St. Bernard,* for the hospice, in 1862. The dog's name happens to be appropriate here, too, since the name *Bernard* literally means "strong as a bear." See also Appendix I.

BASKET was the white poodle owned by the American writer Gertrude Stein (1874–1946). He was so named by Alice B. Toklas, Stein's companion, because as a puppy the dog looked as though "he should carry a basket of flowers in his mouth." He never did, however.

BEAU was the spaniel, "prettiest of his race," owned by the English poet William Cowper (1731–1800). His name obviously suggests a handsome or elegant animal, even a dandy. The name was also that of General Omar Bradley's famous poodle.

BEAUTIFUL JOE is the mongrel whose story is told in the book of the same name (1893) by the American writer Marshall Saunders. He was a real dog, and was so named ironically because his cruel owner had cut off his ears and tail.

BEERBOHM was long the resident cat at the Globe Theatre, London, where he was named for the famous actor-manager, Sir Herbert *Beerbohm* Tree (1853–1917). It has long been considered lucky to have a cat in a theatre. (A well-known fictional one is GUS; see Chapter 12, page 150.)

BELAUD was the silver gray cat who belonged to the 16th-century French poet Joachim du Bellay. He was the subject of his owner's moving *Epitaphe d'un Chat* (circa 1558), and his name is a blend of Old French *bel*, "beautiful" and the poet's own name. Du Bellay also had a pet dog named PELOTON, "ball of wool." Three centuries later, the French novelist Pierre Loti (1850–1923) had a cat BELAUD named for du Bellay's.

BELKA and STRELKA, two Samoyed huskies, were sent into space by the Soviet Union in 1960. Their names are respectively Russian for "Squirrel" (for the dog's appearance) and "Arrow" (for his speed). See also PUSHINKA below.

BENJI was the mongrel puppy who appeared on American television and in movies in the 1960s and 70s, and who played the title role in the film *Benji* (1974). "Ben" is one of the commonest names for dogs, and BENJI is fairly obviously a short form of *Benjamin*.

BILL XXII was the 22nd goat mascot, and the 2nd named BILL, of the football team of the United States Naval Academy, and was acquired in 1979. The first team mascot was EL CID, in 1893, named for the 11th-century Spanish hero of the wars against the Moors, whose own name means "the lord," from Arabic. A later mascot of 1910 was THREE-TO-NOTHING JACK DALTON, named for a football player.

BLACK AGNES was the black palfrey owned by Mary, Queen of Scots (1542–1587). The horse was given to the queen by her brother, Moray, and was named for Agnes Dunbar, Countess of Dunbar and March (1312–1369), nicknamed "Black Agnes" for her dark complexion.

BLACK SALADIN was the coal-black horse owned by Richard Neville, Earl of Warwick (1428–1471). He was named partly for his color and partly for *Saladin,* the 12th-century sultan of Egypt who fought the Crusaders.

BLANCO was the white collie accepted as a gift by American president Lyndon B. Johnson in the 1960s to represent the dozens of dogs that had been offered him. His name actually means "white."

BLEMIE was the Dalmatian belonging to the American dramatist Eugene O'Neill (1888–1953). His name was a short form of EMBLEM, his full official name being SILVERDEEN EMBLEM O'NEILL.

BLONDI was the Alsatian bitch given to Adolf Hitler by his adviser and private secretary, Martin Bormann. Her name means "fair-haired one."

BOATSWAIN was the English poet Lord Byron's favorite dog. He was a Newfoundland, a breed noted for rescuing people from the sea. He was therefore given a "sailor-like" name.

BODGER was the old bull terrier who made his way through the Canadian wilderness with TAO the Siamese cat and LUATH the retriever, as recounted in Sheila Burnford's bestseller, *The Incredible Journey* (1961). The bulldog's name is a dialect term for a traveling dealer or peddler.

BONZO, a chimpanzee, starred in the American movies *Bedtime for Bonzo* (1951) and its sequel *Bonzo Goes to College* (1952). His name is near-arbitrary, but already existed for a dog in a comic strip (see Chapter 2, page 22).

BOUNCE was the great Dane owned by the English poet Alexander Pope. His name describes his lively character and athletic gait. When one of the dog's pups, POLYPHEMUS, was received as a gift by the Prince of Wales, the animal was wearing a collar bearing a couplet by Pope:

    I am his Highness' dog at Kew.
    Pray tell me, Sir, whose dog are you?

BOY was the large white poodle who belonged to Prince Rupert in the English Civil War. The dog, generally believed to have been the first poodle in England, was killed at the battle of Marston Moor (1644). His name is self-descriptive, denoting a dog who is not only male but "ever youthful." The dog was well known in his own lifetime, and is mentioned in contemporary literature, including an anonymous satirical pamphlet in which another dog is mistaken for him (Prince Rupert owned several):

    PRINCE RUPERT'S DOG: What yelping whindling Puppy Dog art thou?
    TOBIE'S DOG: What bauling Shag-haird Cavalliers Dogge art thou?
    PRINCE RUPERT'S DOG: Thou art a dogged sir or Cur, grumble no more, but tell me thy name.

TOBIE'S DOG: I was called *Tobies* house-dog, the Dog which *Walker* the Iron-monger so often commends for a mannerly and well-bred Dog in his severall Tub-lectures; my name is *Pepper.*
PRINCE RUPERT'S DOG: Though your zeal be never so hot, you shall not bite me, *Pepper.*
TOBIE'S DOG: Ile bark before I bite, and talke before I fight, I heare you are Prince *Rupert's* white *Boy.*
PRINCE RUPERT'S DOG: I am none of his white *Boy*, my name is *Puddle.*
TOBIE'S DOG: A dirty name indeed, you are not pure enough for my company.
("A Dialogue, or rather a Parley between Prince Rupert's Dog whose Name is *Puddle,* and *Tobie's* Dog whose Name is *Pepper,* &c.," 1643.)

In this passage, of general interest for its names, Prince Rupert's dog has a name that refers directly to his breed of *poodle*. See this breed name in Appendix I.

BRIGADIER GERARD, the great English racehorse, was foaled in 1968. His name partly refers to that of his sire, QUEEN'S HUSSAR, but it is overall that of the hero of the historical romance *The Exploits of Brigadier Gerard* (1896) by Arthur Conan Doyle.

BRIGHAM was "the best buffalo horse that ever made a track," owned by the famous American showman William F. Cody ("Buffalo Bill") in the 19th century. He was named for Cody's contemporary, American Mormon leader *Brigham* Young.

BROWN JACK, one of the most famous geldings in the history of English racing, was foaled in 1924. His name denotes his color and derives from that of his sire, JACKDAW.

BUCEPHALUS was the black stallion owned in the 4th century B.C. by Alexander the Great. He was the king's favorite steed, and his close companion for 20 years. He is believed to have died of wounds in the fierce battle of 326 B.C. against the rajah Porus and his elephants. His name is of Greek origin and means "ox-headed," from *bous,* "ox," and *kephalē,* "head." This does not describe the horse's own head, but relates to the bull's head with which Thessalian horses were traditionally branded. The steed is said to have worn golden horns in battle as a visual reference to his name.

BUDDY was the first Seeing Eye dog, a German shepherd bitch born in Switzerland. She was originally called KISS when given as a guide dog to Morris Frank, a young Tennessee blind man, but was renamed by him for her companionship. KISS was originally so called because the name began with the letter (K) that identified her particular litter.

BULLET was the German shepherd promoted as Roy Rogers' "wonder dog."

He first appeared in the movie *Spoilers of the Plain* (1951) and later starred on television in *The Roy Rogers Show* (1951-64). His name refers to his speed and strength as an "attacker." Compare the name of TRIGGER, the horse that he accompanied in chase scenes.

BURMESE was the famous black horse presented to Queen Elizabeth of England by the people of Canada in 1969, through the Royal Canadian Mounted Police. She was so named not because she came from Burma but because her lush coat reminded her Canadian trainer of the coat of a *Burmese* cat. The Queen rode BURMESE not only as a hacking horse but annually for the next 18 years at her Birthday Parade (Trooping the Color).

The BYERLY TURK was the Eastern-bred stallion which was one of the three ancestors of all presentday thoroughbreds, the other two being the DARLEY ARABIAN and the GODOLPHIN ARABIAN. He was named for Captain Robert *Byerly*, who captured him at Buda in 1688 when he was fighting the Turks in Hungary.

CAESAR was the wirehaired fox terrier owned by King Edward VII of England (1841-1910). He wore a collar with the inscription: "I am Caesar. I belong to the King," as if to emphasize the connection between his imperial name and the royal title. Soon after the King's death he became widely known as the "author" of *Where's Master?*, a sentimental but also rather touching account, told in the first person, of a sad little dog who searches in vain for his master. The actual author was Sir Ernest Hodder-Williams, chairman of the publishing house Hodder & Stoughton, who published the book. Priced at one shilling, it sold over 100,000 copies within a few weeks of its publication in the year of the King's death.

CALVIN was the large Maltese cat owned by the American essayist and novelist Charles Dudley Warner (1829-1900). He was so named for the famous French theologian and founder of Calvinism, "on account of his gravity, morality, and uprightness." Warner writes of him in his first book of essays, *My Summer in a Garden* (1871).

CARUSO was the canary owned by American president Calvin Coolidge (1872-1933). The bird was a great singer, so was appropriately named for the famous Italian operatic tenor Enrico *Caruso* (1873-1921). A cat of the name was owned by the English writer and critic Edmund Gosse (1849-1928). The purring of a contented cat is sometimes referred to as its "singing."

CATARINA was the cat of the American poet and author Edgar Allan Poe (1811-1849), with a name doubtless based simply on *cat*. Poe adored the cat, who nevertheless was the inspiration for his horror story "The Black

Cat" (1843), in which the cat is renamed PLUTO, for the Greek god of the underworld.

CELER was the horse owned by the 2nd-century A.D. Roman emperor Lucius Aurelius Verus. Its name is the Latin word for "swift."

CHAMPION was the famous horse owned by the American movie star Gene Autry. The actor in fact had several horses of the name. One of the earliest was originally called LINDY, as he was foaled on the day (May 21, 1927) when Charles A. *Lindbergh* flew the Atlantic nonstop from New York to Paris.

CHARLES was the Siamese cat owned for 13 years by the British publisher and cat-lover Michael Joseph (1897–1958). He is the subject of his owner's book *Charles: The Story of a Friendship* (see Bibliography), in which the author explains how he came to give the name, in full CHARLES O'MALLEY:

> It was not my idea. I was lunching with [the writer] Charles Graves, to whom I explained that I wanted something quite unoriental for my amusing, mischievous little cat. "Why not Charles O'Malley, the Irish Dragoon?" he said. So Charles O'Malley (without the Irish Dragoon, for that would have been too fanciful) it was.

The literary reference is to the soldier-hero and narrator of Charles James Lever's popular (and picaresque) novel of Irish life, *Charles O'Malley* (1841).

CHARLEY was the pony ridden by the American explorer, artist and writer George Catlin across 500 miles of prairie in the 1830s for the purposes of studying and painting American Indians. Catlin described him as "a noble animal of the Camanchee wild breed," and his name may have been chosen to reflect this. Another CHARLEY was the French poodle who accompanied the American writer John Steinbeck on his 10,000-mile tour of the United States in 1960. He described his journey in *Travels with Charley* (1963).

CHECKERS was the cocker spaniel who became famous on September 23, 1952, when Senator Richard M. Nixon appeared on American television with the aim of denying that he had improperly accepted funds while running for the Vice-Presidency of the United States. He spoke of his humble background, and in a bid for complete frankness, reported that he had received another gift from an admirer, a cocker spaniel sent in a crate from Texas:

> Black and white spotted. And our little girl—Tricia, the six-year-old—named it Checkers. And, you know, the kids love the dog, and I just want to say this right now—that, regardless of what they say about it, we're gonna keep it!

The dog was named from the fact that its markings suggested the *checkers* on a checkerboard (British: draughts on a draughtboard). The broadcast went down in history as "The Checkers Speech."

CHING-CHING and CHIA-CHIA were the pair of giant pandas, respectively female and male, that were presented by the Chinese government to the London Zoo, England, in 1947. Their respective names mean "very bright," from a reduplication of Chinese *jīng*, "brilliant," "glittering," and "most excellent," "very good," in similar fashion from *jiā*, "good," "fine."

CHINKAPEN was a dog owned by American president Lyndon B. Johnson (1908–1973). The dog was a small one, and his name was actually the nickname of the president's daughter Luci, known for her small size as "Luci Chinkapen." A *chinkapen* is a dwarf chestnut. Luci herself had beagles named PECOSA and ASTRONAUT. The first of these is Spanish for "freckled." The second name was a compliment to the American astronauts who were by then (1960s) beginning to be in the popular eye when training for the famous Apollo project of 1969 (the year that L.B.J. was succeeded by President Nixon).

CHOLMONDELEY was the chimpanzee shipped from Africa to the London Zoo by the British zoologist Gerald Durrell, who described him in *The Overloaded Ark* (1953). His name, pronounced "Chumley," is a blend of the traditional English aristocratic surname and *chum*, with perhaps also a hint of *chimp*.

CINCINNATI was the horse who was a gift from the people of the Ohio city of this name to General Ulysses S. Grant. He was his owner's favorite mount, and was taken to Washington in 1869 when Grant became president.

COCO was the French poodle owned by the veteran English film critic Dilys Powell. In *Coco: A Biography* (1952), his mistress tells how she and her husband came to choose the name. They at first tentatively gave the pup a French name: MONSIEUR TOUTOU, "Mr. Bow-wow," having heard this name used in France by a ticket collector for a dog on a train. They then chose PLUTO, as his "enthusiastic lolloping" reminded them of the Disney dog. His brown color and crazy behavior final settled them in favor of COCO: "cocoa by colour, Coco by nature: Coco the clown." Miss Powell, a writer on Greece as well as a movie critic, had earlier owned a Siamese cat named PERIANDER (PERRY for short), "after the tyrant of Corinth."

COFFEEBEAN was the nanny goat who was the childhood pet of the English soldier, explorer, and zoologist Ronald Kaulback (born 1909). Her full name was CICELY AUDREY COFFEEBEAN KAULBACK: "I cannot now remember why we called her Cicely, but Audrey was the name of our governess, a great favourite of ours; and, as for Coffeebean—well, there should be no need to explain that one." (In case there is: her milk was used for coffee!)

COLO was the first gorilla born in captivity, at the Columbus, Ohio, zoo on December 22, 1956. Her name is a pet form of *Columbus*.

COMANCHE was the horse of Captain Myles Keogh at the battle of Little Big Horn, Montana (1876). As a military mount, he was the sole survivor of "Custer's last stand" in this fierce fight against the Indians, in which Custer himself and all his forces were killed. Ironically, his name is of Indian origin, and derives from the *Comanche* Indians who formerly ranged from the Platte River to the Mexican border. Their own name is said to mean "bald-heads," from their custom of shaving their heads.

COPENHAGEN was the Duke of Wellington's famous charger, bought by the military leader in 1812 for the Peninsular Campaign. The Duke rode him at the battle of Waterloo (1815). The horse died in 1836 and was buried with military honors at the Duke's estate of Stratfield Saye, Hampshire. He was named commemoratively for the Duke's part in the *Copenhagen* expedition of 1807. Compare the name of MARENGO, below.

DALAL is the elephant who was one of the few animals to survive in Kuwait Zoo when the country was invaded by Iraq in 1990. Most of the 738 animals were killed in the occupation, or were slaughtered to feed soldiers. Only 35 survived, including DALAL, who was the "senior resident" when the Zoo was reopened in 1992. Her name is Arabic, and means "coquetry." The biblical name *Delilah* may be related.

The DARLEY ARABIAN was the bay Eastern-bred stallion which, together with the BYERLY TURK and the GODOLPHIN ARABIAN, was one of the three ancestors of all presentday thoroughbreds. He was foaled in 1870 and bought from Sheik Mizra III by Thomas *Darley,* Queen Anne's consul in Aleppo, now in Syria. See also LEXINGTON (below).

DASH has been the name of more than one well-known dog, including the "large and very handsome dog, of a rather curious breed" owned by the English poet Thomas Hood (who passed him on to Charles Lamb) and, more famously, the King Charles spaniel of Princess Victoria, who was given him by her mother, the Duchess of Kent, in 1833, when she was 13 years old. The Princess became Queen Victoria in 1837 and moved to Buckingham Palace, when she noted in her diary that "dear *Dashy*" loved the garden. The dog died in 1840 and was buried at Windsor Park with the inscription:

READER
If you would be beloved and regretted
Profit by the example of
DASH

The name either implies a lively nature or showy appearance, that of an animal which "dashes about" or "cuts a dash," or else describes a coat marked as if "dashed" with different colors. Lamb constantly complained that the dog scampered up and down "all manner of streets" when out for a walk, which suggests that his name was more than appropriate.

The association of the name with a dog that disappears is alluded to by Lewis Carroll in *Through the Looking-Glass* (1871), when Alice wonders what will become of her name when she enters the wood. Maybe one of the creatures she finds there will have it? "That's just like the advertisements, you know, when people lose dogs— *'answers to the name of "Dash": had on a brass collar'.*"

DAWN RUN, foaled in 1978, was the Irish mare who gained fame as the first horse to win both the important Cheltenham races, the Champion Hurdle and Gold Cup. Her name derives from those of her sire and dam, respectively DEEP RUN and TWILIGHT SLAVE. A "dawn run" is also the daily early morning gallop taken by horses when training.

DESERT ORCHID, one of the most popular English steeplechasers of recent times, was foaled in 1979. His sire was GREY MIRAGE and his dam FLOWER CHILD. To the racegoing and betting public he was familiar as DES or DESSIE. When young, he had an unconventional appearance. His groom therefore initially called him by the most conventional name she could think of, FRED. For professional purposes, however, he needed a professional name. His owner has described how the name evolved:

> We needed to give him a name. We went to the stable and looked at Fred. He looked back blankly. No help there. Then a famous poem came to mind, Thomas Grey's *Elegy Written in a Country Churchyard,* and the passage "Full many a flower is born to blush unseen,/ And waste its sweetness on the desert air." [ . . . ] We couldn't have the name Desert Air. Some bright spark at Weatherbys [publishers of registered names of horses] had noticed there was already a Desert Heir, and from a commentating point of view this could be confusing in the unlikely event of both horses being in the same race. Back to square one. We all agreed that no good horse ever had a bad name. "Let's keep the 'Desert' anyway," I said. We agreed. There was a pause. "I really loved Grey Orchid, you know," said Jimmy [the horses' breeder, and the owner's father], "and she *was* the start of all this" (Grey Orchid was Des's grandmother). We thought about it for a moment. Desert Orchid. It seemed perfect, and the name was available [Richard Burridge, *The Grey Horse: The True Story of Desert Orchid,* 1992].

## 10. Famous Real Animals

DIOMED, foaled in 1777, is famous as the winner of the first Derby, in 1780. He was probably named for the mythological *Diomedes* who was king of Argos and fought with the Greeks at Troy, rather than for the equally mythological *Diomed* who fed his horses on human flesh.

DUKE was the puppy given to the French philosopher and writer Jean-Jacques Rousseau, who said of him in his *Confessions* (1764–70) that he "certainly deserved the name better than the majority of those who had assumed it." Rousseau later renamed the dog as TURK, and was embarrassed when an aristocratic acquaintance obliged him to admit, in the presence of several dukes, that the dog was originally called DUKE.

ECLIPSE, one of the greatest English racehorses of all time, was foaled on April 1, 1764, during a total eclipse of the sun. Hence his name.

ELSA became famous as the lioness rescued as a cub by George Adamson, a game warden in Kenya. She formed the subject of two popular books by George's wife, Joy Adamson, *Born Free* (1960) and *Living Free* (1961), and was made known to a wide public through the movie *Born Free* (1966) and a subsequent television series. Joy Adamson tells us that she named the cub for "someone she knew." This was actually her mother-in-law, *Elsa* von Klarwill.

ENJOLRAS, GAVROCHE and EPONINE were a trio of coal-black cats owned by the French poet and critic Théophile Gautier (1811–1872). They were all named for characters in Victor Hugo's novel *Les Misérables* (1862), respectively suggesting the greediness of Enjolras, the "street Arab" nature of Gavroche, and the love expressed by Éponine (in the novel, for Marius). (Hugo himself had a cat named GAVROCHE; the animal was also known as CHANOINE, literally "canon," but with specific reference to the French expression *mener une vie de chanoine*, "lead an easy life.") Gautier and his daughter Judith were great cat lovers. Other cats owned by the father included the shorthair CHILDEBRAND, named for an 8th-century Frankish prince; MADAME-THÉOPHILE, so named because she lived with her master "on a footing of conjugal intimacy"; DON PIERROT DE NAVARRE, originally named PIERROT for his white color, like the costume and face of a *Pierrot* or clown, but later acquiring a longer name which suggested a Spanish grandee; and SÉRAPHITA, an "aristocratic" cat with a poetic name. Those owned by Judith Gautier (1850–1917) included the black cat SATAN, the small cat BÉBÉ, and others named LILITH (for an Assyrian demon), IBLIS (for the chief devil in Islamic mythology), and CREVETTE ("shrimp").

FALA was the Scottish terrier owned by American president Franklin D. Roosevelt (1882–1945). As a pup he was called BIG BOY, but was then renamed half-jokingly as MURRAY THE OUTLAW OF FALA HILL, or FALA for short. *Fala* Hill is near Edinburgh, Scotland, and the renaming was made by Roosevelt as a tribute to a remote Scottish ancestor. FALA mated with a Scottie named BUTTONS, and the offspring were two bitch puppies MEGGY and PEGGY, both Scottish-style names. An entirely appropriate name was that of PRESIDENT, a great Dane presented to Roosevelt.

FELLER was a dog owned by American president Harry S Truman (1884–1972). The name simply means "fellow." Truman's daughter Margaret had an Irish setter named MIKE, a name associated with the Irish (although more commonly in the form *Micky*).

FIDO was the mongrel who belonged to Abraham Lincoln's two younger sons, Willie and Tad, in Springfield, Illinois. His name is a long traditional one for a faithful dog (see Chapter 2, page 21).

FLOSSY was the small black and white spaniel owned by the English author Anne Brontë (1820–1849). The name is strictly speaking a pet form of *Florence*. For a pet dog, however, it suggests *flossy,* either in the sense "downy," "having soft hair," or as "showily attractive," "flashy," or even a blend of both.

FLUSH was the red cocker spaniel who was the pet of the English poet Elizabeth Barrett Browning (1806–1861). Virginia Woolf's *Flush* (1933) is a biography of the dog from his point of view. The name is appropriate for a cocker spaniel, used as a sporting dog to *flush* game.

FLYING CHILDERS, the first great racehorse of the English turf, was foaled in 1714 and named for his Yorkshire breeder, Leonard *Childers,* of Carr House, near Doncaster.

FOINAVON, winner of the Grand National in 1967, was named for the Scottish mountain of *Foinaven* (as now usually spelled), west of Inverness. For a racehorse with a similar name, compare ARKLE (above).

FOSS was the striped cat who was the companion of the English nonsense poet Edward Lear (1812–1888). The name is a shortening of Greek *adelphos,* "brother," as the cat was the brother of an earlier cat owned by Lear that had disappeared.

FURY was the black stallion who starred in the American television series named for him and who also appeared in movies such as *Black Beauty* (1946), *Gypsy Colt* (1955), and *Wild in the Wind* (1958). The name is almost traditional for a spirited horse.

GANDHI was the dachshund given by the American newspaper tycoon William Randolph Hearst to his protégée, the actress Marion Davies (1897–1961). The dog's full whimsical name was MAHATMA COATMA COLLAR GANDHI.

GARGANTUA was the gorilla who was the central attraction of Ringling Brothers and Barnum and Bailey Circus in the 1940s. His name seems to suggest "gigantic," but is actually a borrowing of that of the fictional character *Gargantua* who is the giant hero in Rabelais' 16th-century satire *Gargantua and Pantagruel*. The name represents Spanish *garganta*, "throat," "gullet," so is suitable for an animal with a voracious appetite.

GATO and MANCHA were the small, hardy Criollo horses who accompanied the Swiss-born schoolmaster, A.F. Tschiffely, on a 13,000-mile ride from Buenos Aires to Washington, D.C., in 1925. Their names are Spanish, and mean respectively "cat" and "spot"; GATO was light brown with black feline streaks; MANCHA was a pinto, or skewbald, with variegated patches of white and red.

GEIST was a dachshund owned by the English poet Matthew Arnold (1822–1888). His German name means "ghost," "spirit."

GEORGE was the tallest giraffe ever kept in captivity (in Chester Zoological Gardens, England). His name is an arbitrary one designed to alliterate with *giraffe*.

GIMCRACK, foaled in 1760, was one of the best known English racehorses of the 18th century. His name represents the standard word *gimcrack*, meaning "dodge," "trick," but also suggests a name "Jim Crack," for a horse who runs at a cracking pace.

GIP was a large tabby owned by the American natural history writer W.H. Hudson (1841–1922). His name was presumably a short form of GIPSY.

The GODOLPHIN ARABIAN or GODOLPHIN BARB was the bay stallion, foaled in Yemen in 1724, which together with the BYERLY TURK and the DARLEY ARABIAN, was the ancestor of all thoroughbred horses today. He is named for Francis *Godolphin,* 2nd Earl of Godolphin (1678–1766), who acquired him some time after 1731. His alternate name derives from the fact that he had earlier been exported to Tunis, on the Barbary Coast in North Africa.

GOLDEN MILLER, foaled in 1927, was a gelding bred in Ireland who went on to win five Gold Cups in succession as well as the Grand National. His name comes from those of his sire and dam, respectively GOLDCOURT and MILLER'S PRIDE.

GOLDIE was the eagle whose escape from the London Zoo, England, on February 27, 1965, excited much media attention. The bird was originally unnamed:

> The eagle came to be called "Goldie." This name had to be hastily thought up for the inquiring press by one of the Zoo's officials, as the bird population at the Zoo is too numerous for them all to have individual pet names. But it was a good name, and in spite of—or because of—its banality, the British took it to their hearts at once. (Jonquil Antony, *Eaglemania*, 1965.)

GOLDSMITH MAID was "the trottin'st mare in history," foaled in New Jersey in 1857. When she began to race in 1865 she was named LADY GOLDSMITH, but was later given the altered name by which she is still remembered today. Either way, it commemorates her owner, Alden *Goldsmith*, on whose farm near Goshen, New York, she grew up.

GOMA was the second gorilla to be born in captivity (after COLO, above). She was born at the Basel Zoo, Switzerland, in 1959 and was given a Swahili name (for the animal's native West Africa) meaning "Dance of Joy."

GREYFRIARS BOBBY is famous in Scottish folklore as the Skye terrier who kept a nightly watch for 14 years on the grave of his master, a shepherd named John Gray, known as "Auld Jock." The dog takes his name from the *Greyfriars* Churchyard, Edinburgh, where the old man was buried, and where in the course of time he would also lie.

GREYHOUND was one of America's greatest trotters, foaled in 1932. He was named for his color and speed, which also gave his nickname, "The Gray Ghost."

GRITS was the only dog, a mongrel, owned by American president Jimmy Carter (born 1924), or more precisely by his daughter Amy (born 1967). The name was appropriate for a pet whose master, the son of a grocer and peanut broker, had made it from humble origins in the Deep South to the pinnacle of national government.

GUY was the popular gorilla who arrived at the London Zoo, England, on November 5, 1947. He was named for this date, *Guy* Fawkes Day.

HACHIKO was an Akita dog who accompanied his master to the Shibuya railroad station, Tokyo, every morning and who waited for him to return every evening. One day in 1925 his owner died suddenly at work. In the evening the dog went to the station as usual, but when his master did not return, came back the following evening and repeated his visit every day after for almost nine years. His name means "eighth child," as he was the eighth pup of a litter.

HAM, known as the "Space Chimp," was the American chimpanzee sent into space in 1961. He was originally known simply as "test subject Number 61," but was then named HAM as an acronym for *H*olloman *A*erospace *M*edical Center, the scientific establishment that prepared him for the flight. The chimp subsequently became a celebrity at the National Zoological Park in Washington, D.C.

HAMBLETONIAN, foaled in 1849 in Orange County, New York, was the most famous trotting stallion in America, and the ancestor of most present harness-racers. He was named for a well-known English thoroughbred, winner of the St. Leger in 1795.

HEIDI was a Weimaraner owned by American president Dwight D. Eisenhower (1890–1969). The name is a typically German one, to match the breed. Eisenhower's grandson David had a lively little Scottie called SPUNKY, who was no friend of the larger dog.

HEROD, originally KING HEROD, named for the biblical ruler, was a famous English thoroughbred racehorse foaled in 1758. Together with ECLIPSE and MATCHEM he was one of the three principal progenitors of all thoroughbreds now racing.

HIM and HER were a pair of beagles owned by American president Lyndon Johnson in the 1960s. Their unoriginal names simply indicated their sex.

HINSE OF HINSFIELD was the family cat of the famous poet and novelist Sir Walter Scott (1771–1832). He was named for the hero of a German folktale, but with the name assimilated to a fictional place in the Scottish manner, as for the mountain Cairnsmore of Fleet, for example.

HODGE was famous as the cat of the English lexicographer Dr. Samuel Johnson (1709–1784). *Hodge* was long a traditional nickname for an English agricultural laborer or rustic, and Johnson may have named his cat with this in mind. On the other hand, he may have regarded his pet as something of a *hodge-podge,* a term defined by Johnson himself in his famous *Dictionary of the English Language* (1755) as "a medley of ingredients boiled together."

HUMPHREY is the black and white tomcat who is the official mouser at 10 Downing Street, London, the Prime Minister's residence. He was appointed after the death of the long-serving WILBERFORCE in 1988, and is named for Cabinet Secretary Sir *Humphrey* Appleby, a leading character (played by Nigel Hawthorne) in the television series *Yes, Prime Minister* (1986–7).

HURLYBURLYBUSS was a cat owned by the English poet Robert Southey (1774–1843). His name, stated by his owner to be "of Germanic and Grimmish extraction," is a fanciful classical embellishment of *hurlyburly,* but with a nod in the direction of German *hurliburli,* "precipitately." Southey also had RUMPEL (see below).

HYPERION, the famous English racehorse foaled in 1930, has a name that rather perversely refers to that of his dam, SELENE. In Greek mythology, *Selene* is the goddess of the moon who is the *daughter* of the Titan god *Hyperion.* Although he was not a steeplechaser, HYPERION's name is literally appropriate for a racehorse, since in the original Greek it means "going over" or "high walker."

IGLOO was the fox terrier who accompanied the American explorer Admiral Richard E. Byrd on his first Antarctic expedition of 1928–30. He was named for the familiar Eskimo house, made of snow blocks.

INCITATUS was the favorite horse of Caligula, the eccentric Roman emperor of the 1st century A.D. The horse was given a marble stable and a fully furnished house in which slaves waited on the guests whom his master invited to dine with him. A legendary story tells how the emperor planned to make his steed a consul. The horse's name means "incited," "spurred on," from Latin *incitare,* "to urge forwards."

IROQUOIS gained fame as the first American-bred horse to win the English Derby, in 1881. He was named not for his sire or dam (respectively LEAMINGTON and MAGGIE B.B.) but for the well-known North American Indian people, whose own name is said to mean "worst enemies," as distinct from the Sioux, "least enemies."

J. FRED MUGGS was a chimpanzee familiar from his appearances in the "Today" show on American television from the 1950s. He was given an arbitrarily conventional name, with *Muggs* implying *mugging,* the actor's art of grimacing or making exaggerated gestures for comic effect.

JOCK was the cat given to Sir Winston Churchill on November 30, 1962, the prime minister's 88th birthday. He was named for his donor, Sir Winston's friend and former Private Secretary, Sir John ("Jock") Colville. He died in 1975, and his successor, also named JOCK, was provided for in Sir Winston's will, which stipulated that a marmalade cat should be in "comfortable residence" at his Chartwell home for ever. JOCK II died in 1991 at the ripe feline age of 16. JOCK III was duly selected in 1992, after an exhaustive search for a cat whose looks and coloring most closely matched those of the original.

JUMBO is generally regarded as the world's most famous elephant, and was probably the largest one ever kept in captivity. (Hence *jumbo* for any large object, such as "jumbo jet," "jumbo crossword," and the like.) He was acquired in 1865 by the London Zoo, England, and in 1882 sold to the American showman P.T. Barnum. His name is generally said to represent the second word of *mumbo jumbo,* itself originating from the name of an African tribal god. However, the following letter appeared in the winter 1948 number of the magazine *Zoo Life:*

> THE ORIGIN OF THE NAME "JUMBO"
> The word "Jumbo" has so completely entered into our language that it is included in the dictionaries, but even the Oxford Dictionary does not make any suggestion about its origin. I submit, therefore, the following, which seems to me very probable. In several Bantu languages the word for elephant is *Njamba,* or something very close to it. That is the case in Chokwe or Chibokwe, and readers of Livingstone may remember that he had trouble with a chief of that name. I had a gun boy in Angola, of the same tribe, also of that name, and I believe the word is the same

in the Umbundu group of languages. As Jumbo was an African elephant, this seems to me to be a very probable origin of the name. Malcolm Burr.

JUSTIN MORGAN is unique in being the only horse for whom a distinct new breed was named. For the origin of his name, see *Morgan* in Appendix I.

KANTHAKA was the favorite stallion of the Buddha. The horse had been formerly ridden into battle by the Buddha's father, and was returned to him when the Buddha renounced his worldly life at the age of 29 in the 6th century B.C. to seek enlightenment as an ascetic. The name of the horse is the Sanskrit word for "necklace," perhaps referring to markings on the horse's neck or to an actual necklace that it wore.

KEEPER was the bull mastiff owned by the English writer Emily Brontë (1818–1848). His name was a common one for dogs in the 18th and 19th centuries, and was suitable for an animal that is a "guardian," "one who keeps watch."

KINCSEM, foaled in 1874, was one of the most successful Hungarian racehorses, running in many European races. Her name is Hungarian for "my treasure."

KING TIMAHOE was the Irish setter puppy presented to American president Richard M. Nixon by his New York campaign staff in January 1969. The president named him for "the little village in Ireland where my mother's Quaker ancestors came from," and added *King* because "even a President's dog gets the royal treatment." Not surprisingly, the dog's pet name was simply TIM.

KING TUT was the police dog acquired in World War I by American president-to-be Herbert Hoover (1874–1964). The dog bears the colloquial name of the ancient Egyptian king Tutankhamen, whose tomb was discovered in 1922. Hoover's other dogs included BIG BEN, a prize-winning fox terrier, GLEN, an aptly named Scottish collie, WEEJIE, a small elkhound, and YUKON, an Eskimo dog, named for the river and territory associated with Eskimos.

KOKO, the first gorilla to learn sign language, was taught the hand signs of the American Sign Language for the deaf in the 1970s. Her name, commonly found for apes and monkeys, probably derives from the same Spanish word, *coco*, "bogyman," "face," that lies behind the *coco*nut.

LAIKA was the first living creature to orbit the earth. The Samoyed husky bitch was sent into space by the Russians in 1957. Her name is properly that of her breed, and literally means "barker." See the breed name in Appendix I for details. The dog's pet names were KUDRYAVKA, "curly," for her coat, and LIMONCHIK, "little lemon," for her color.

LEXINGTON was the greatest thoroughbred foaled in the United States in the 19th century. He was originally called DARLEY, for the famous DARLEY ARABIAN (see above). The horse was renamed LEXINGTON for the Kentucky city where he raced (and which is itself now famous for its breeding of thoroughbreds).

LIEN-HO was the baby giant panda presented by the Chinese government to the London Zoo, England, in 1946. Her name represents Chinese *liánhé*, "union," as she symbolized friendship between Britain and China.

LOOTIE, a fawn and white Pekinese, was the first of her breed to be taken to the west from China, when the Summer Palace in Peking was *looted* or ransacked in 1860. She was presented to Queen Victoria by General Sir John Dunne, who was present when Peking was entered by French and English troops that year. See also *Pekinese* in Appendix I.

LUATH was the favorite dog of the Scottish poet Robert Burns (1759–1796). He appears as the ploughman's rough collie, the "gash an' faithfu' tyke," in Burns' satirical poem *Twa Dogs* (1786), where he wanders round with his friend CAESAR, the dog who lives with the gentry. A hundred years later, LUATH was the black and white Newfoundland dog who belonged to the English author J.M. Barrie (1860–1937) and who was the model for NANA (see Chapter 12, page 155) in the play *Peter Pan* (1904). Both these dogs were named for the legendary Irish dog LUATH, although Burns' poem gives the source of the name rather differently:

> An' in his freaks had Luath ca'd him,
> After some dog in Highland sang,
> Was made lang syne, Lord knows how lang.

(See Chapter 11, page 135, for the origin of the name.) More recently, LUATH was the Labrador retriever who accompanied the bulldog BODGER and Siamese cat TAO in Sheila Burnford's *The Incredible Journey* (1961).

MACARONI was the piebald pony who in the 1960s belonged to Caroline Kennedy, daughter of American president John F. Kennedy. He was so named for his mixed (*"macaronic"*) coloring. The name also rhymes with *pony*, and even suggests that of his owner.

MAIDA, a cross between a wolfhound and a deerhound, was a favorite dog of the Scottish author Sir Walter Scott. He was given to the Scottish writer by his friend Glengarry, who had named him "out of respect for that action in which my brother had the honour to lead the 78th Highlanders to victory," i.e., the Battle of *Maida* in 1806, when the English defeated the French in the Napoleonic Wars near Maida, in Spain. MAIDA was the model for Scott's fictional dogs BEVIS and ROSWAL (see Chapter 12, pages 143 and 159).

MAN O'WAR, the most famous thoroughbred racehorse in America, was foaled in 1917 and was named (originally as MY MAN O'WAR) by Eleanor Belmont, wife of his owner, Major August Belmont, who at the time was engaged in active military service. He was known to the public as "Big Red," for his golden chestnut color.

MARENGO was Napoleon's favorite charger, a white Arabian stallion, ridden by his master at the battle of Waterloo (1815). He died in England, where he had been taken as a prize of war, in 1829. The horse's name, like that of COPENHAGEN (see above), is commemorative, for the battle of *Marengo* (1800), Italy, in which Napoleon won a narrow victory in the War of the Second Coalition.

MARGATE was the black kitten owned by Sir Winston Churchill. It turned up on the step of 10 Downing Street on October 10, 1953, the day of the prime minister's important speech to the Conservative Party at *Margate, Kent*. As a token of good luck, it was thus named for that town.

MAROCCO was the famous performing horse of Elizabethan times in England, owned and trained by one Thomas Bankes or Banks, a Scottish showman. The horse's name suggests that he was a Barb, but proof of his precise origin and ancestry is lacking.

MASTER MAGRATH was the famous greyhound racer owned by Lord Lurgan in the 1860s, and winner of the Waterloo Cup three years in succession. The dog is said to have been rescued by Lurgan when about to be drowned as a pup, and to have been named for the Irish orphan boy who raised him.

MATCHEM has been called the first thoroughbred. He was foaled in England in 1748 as the grandson of the Godolphin Arabian, and was one of the three principal progenitors (with ECLIPSE and HEROD) of most modern thoroughbreds. His name implies that he can hold his own with any other horse, that he can "match'em."

MATHE, the favorite greyhound of Richard II of England, has a name that was a medieval spelling of *Matthew*. It is uncertain to what extent the king related the name to its literal meaning, "gift of God."

MIDNIGHT was a champion buck-jumper of the Canadian and American rodeo. He was born in 1916 in Alberta, Canada, and was named for his dark, shining coat. He is buried near Plattsville, Colorado, where an inscription over his grave was written by the cowboys who loved him:

> Under this sod lies a great bucking hoss,
> There never lived a cowboy he couldn't toss,
> His name was Midnight, his coat black as coal,
> If there is a hoss-heaven, please God rest his soul.

MIJBIL is the otter who is effectively the central character of Gavin Maxwell's *Ring of Bright Water* (1960), the author's autobiographical account of his simple life on the northwest coast of Scotland, with the otter as a pet. Maxwell had obtained the otter on a visit to Iraq in 1956, and named him for a sheikh with whom he had been staying. The name intrigued him, as it seemed to suggest a platypus-like creature.

MILL REEF, the famous English middle-distance racer, was foaled in 1968 and derived his name from those of his sire and dam, respectively NEVER BEND and MILAN MILL. Both "reef" and "bend" are kinds of knot.

MILLIE is the English springer spaniel owned by American president George Bush and his wife Barbara. The dog hit the headlines in 1989 when her puppies were born, and her "autobiography," *Millie's Book*, "by Millie" (in reality by Barbara Bush), was a bestseller when it appeared the following year. Barbara Bush explained in a magazine article: "Millie is named after a very close friend of mine, Mildred Kerr—that is *not* c-u-r, but Kerr" (*Life*, May 1989, page 34).

MINNA MINNA MOWBRAY was a favorite cat of the English publisher and cat-lover Michael Joseph (1897–1958). Although the cat features in his book *Cat's Company* (see Bibliography), his owner does not give the source of the name, which nevertheless strongly suggests the French conventional *min-min-minou* that corresponds to the English PUSS-PUSS-PUSS used to call a cat whose name is unknown. (See also the name MINOU in Chapter 2, page 20.)

MR. FING, described by New York journalist Zoe Beckley as "a jet-black, fat-jowelled, steady-eyed tomcat," was the pet of the theatre and cinema fashion designer Gordon Conway, who took him everywhere with her. He started life as a Soho, London, alley cat called BILLY, but when he became part of Gordon Conway's life as an international celebrity, she gave him a new name that had "a good ring to it" and that suggested he was of Oriental origin. He died in 1934, aged 13.

MODESTINE was the "diminutive she-ass, not much bigger than a dog" with whom Robert Louis Stevenson traveled through France, as recounted by him in *Travels with a Donkey in the Cévennes* (1879). He named the donkey ("Modestine, as I instantly baptised her") for her small size and for the modest amount he paid for her ("sixty-five francs and a glass of brandy"). Her name also happens to echo that of the town where Stevenson bought her, *Le Monastier*.

MORZILLO or EL MORZILLO was the favorite black stallion of the Spanish conquistador Hernando Cortés, taken with other horses from Cuba to Mexico in 1519. His name means "the black one," and corresponds to modern Spanish *morcillo,* a black-coated horse with a reddish sheen.

NELSON, George Washington's favorite horse for cross-country hunting, was a light chestnut gelding. The president named him for his donor in 1765, Thomas *Nelson,* governor of Virginia.

NIJINSKY, the well-known Irish racehorse, foaled in 1967, had a name that came directly from the famous Russian ballet dancer Vaslav *Nijinsky.* This was itself suggested by the name of his sire, NORTHERN DANCER (see below). Nijinsky began his dance career in St. Petersburg, in the north of Russia.

NIM CHIMPSKY was a chimpanzee taught to use sign language in an experiment devised at Columbia University in 1973. The aim was to see if he could create sentences innately, as a child does. His name is thus a punning alteration of that of *Noam Chomsky,* the American linguist who claimed that this ability was peculiar to humans.

NIPPER was the black and white fox terrier who was the model for the dog listening to the gramophone on the British record label *His Master's Voice.* The name is a common one for a small, active dog, who *nips* in and out of restricted spaces, and who also bites or *nips.*

NORTHERN DANCER, the greatest racehorse to come out of Canada, was foaled in 1961. His name partly relates to his country of origin, and partly to that of his sire, NEARCTIC.

OLD ROWLEY was a favorite horse of Charles II of England. The origin of his name is uncertain. The king was himself known as "Old Rowley," allegedly because he was frequently seen riding this horse. However, some derive his nickname from the phrase "a Roland for an Oliver," referring to the fact that Charles (as "Roland") succeeded *Oliver* Cromwell as ruler of Great Britain. The *Rowley* Mile, a stretch of the racecourse at Newmarket, is presumably named for the horse.

OSHKOSH was a white collie presented to American president Calvin Coolidge (1872–1933) in 1923. The dog was so named simply because he came from this Wisconsin town. Mrs. Coolidge, however, later changed the name to ROB ROY, as more appropriate for a Scottish breed, despite the fact that this name itself means "red," not "white." Another collie owned by the president included PRUDENCE, familiarly known as PRUDY (or, by Mrs. Coolidge, PRUDENCE PRIM.) The President's wife also owned a small red-haired chow called TINY TIM. He later grew fat, and was nicknamed TERRIBLE TIM.

Coolidge and his wife were devoted dog lovers, and others owned by them included PETER PAN, a troublesome terrier, DIANA OF WILDWOOD, a Shetland sheepdog (renamed CALAMITY JANE by Mrs. Coolidge for her proneness to accidents), a chow BLACKBERRY, a bulldog BOSTON BEANS, another collie FOXY (properly RUBY ROUGH, but nicknamed MULE

EARS), and KING COLE, a black Belgian police dog, his name punning on his color.

PERSIAN WAR, the English bay gelding who became a champion handicap hurdler, was foaled in 1963 and named for his sire and dam, respectively PERSIAN GULF and WARNING. The Persian Wars themselves were those fought by Greece in the 5th century B.C.

PERSIMMON, owned by Albert Edward, Prince of Wales, was the racehorse who gained fame when he won the Derby in 1896 at odds of 5–1. His name derives from that of his sire, ST. SIMON, so that he was "per ST. SIMON," or "by ST. SIMON." The name itself is also a word in its own right for a type of date-plum tree.

PETITE ETOILE, the English gray filly owned by Prince Ali Khan and his son, the Aga Khan, was foaled in 1956. Her name, French for "little star," punningly comes from those of her sire, PETITION, and her dam, STAR OF IRAN.

PETRA was the dog who was the first of a long line of animals to appear on the British children's television program *Blue Peter*. She arrived in 1962, and was regularly present until 1977. Her name was chosen from around 10,000 proposals sent in by young viewers. It obviously relates to the program title itself, while also suggesting *pet*. It later tied in well with the name of the program's presenter from 1967 to 1979, *Peter* Purves.

PHALLUS was the favorite steed of the 7th-century A.D. Byzantine emperor Heraclius. His name means simply "stallion," from Greek *phallos*, "male sexual organ."

PHAR LAP, the famous Australian racehorse, was foaled in 1926 and at first sight appears to have a name intended to suggest "far lap." It is actually a Thai name, meaning "flash of lightning."

PHRENICOS was the horse owned by the 5th-century B.C. ruler of Syracuse, Hieron I. The horse won the prize for single horses in the 73rd Olympiad (488 B.C.). His name, appropriately, means "intelligent," from Greek *phrēn*, genitive *phrenos*, "sense," "instinct."

PUSHINKA was a Russian laika presented to Jacqueline Kennedy, wife of American president John F. Kennedy (1917–1963), by Soviet premier Nikita Khrushchev. The dog's name means "Fluffy," and she was the daughter of the famous "space dog" STRELKA (see BELKA, above). The bitch later produced offspring "pupniks" with straightforward English names: BLACKIE, BUTTERFLY, STREAKER and WHITE TIP. However, the name of STREAKER appears to suggest that of his grandmother, STRELKA, and by a linguistic coincidence has almost exactly the same meaning. J.F.K.'s father, Joseph Kennedy, presented his daughter-in-law with a German shepherd called CLIPPER, while SHANNON was an Irish cocker (named for the Irish river and airport) offered to the Kennedys as a gift from the Irish prime minister, Sean F. Lemass. Another Irish present was an Irish wolfhound, rather unenterprisingly named WOLF. The president's brother, Senator Robert F. Kennedy, owned a beagle named FRECKLES. On the president's death, this dog passed to his successor in the White House, Lyndon B. Johnson (see HIM and HER, above).

QUEENIE was the Alsatian (German shepherd) owned by the British writer and editor J.R. Ackerley (1896–1967). She was renamed TULIP for the purposes of his autobiographical account *My Dog Tulip* (1956), in which he describes his intense relationship with her, and she also has this name as the subject of his novel *We Think the World of You* (1960). The name itself is found for other animals, and is generally intended to suggest the delicate spring flowers. Ackerley dedicated a later autobiographical work, *My Father and Myself* (1968), to QUEENIE, but her name was altered posthumously to TULIP to accord with the earlier book.

RAB is the huge mastiff who is one of the main canine characters in Dr. John Brown's highly popular book *Rab and His Friends* (1859). Brown was a Scottish physician, and the dog, with his typically Scottish name (a diminutive form of "Robert"), was one he had befriended as a boy.

RACAN and PERRUQUE were two of the many cats owned by the French minister of state, Cardinal Richelieu (1585–1642). These two were named from the fact that they were both born in the wig (French *perruque*) of the poet Honorat de Bueil, Seigneur de *Racan*, a friend of the cardinal. Other cats owned by the cardinal included his favorite, SOUMISE ("submissive"), LUDOVIC LE CRUEL (so named for his cruelty to rats), LUCIFER (for his jet black color), and PYRAME and THISBÉ ("Pyramus and Thisbe," so named because, like the famous lovers, they slept in each other's arms).

RAIN LOVER, the Australian bay colt who won the Melbourne Cup twice, in 1968 and 1969, was foaled in 1964 and takes his name from his parents. His sire was LATIN LOVER and his dam RAIN SPOT.

RECKLESS was the Korean racing mare who was trained to carry ammunition for a U.S. Marine Recoilless Rifle Platoon in the Korean War of the early 1950s. Her name is a form of *Recoilless*.

RED RUM, the English winner of three Grand Nationals, was foaled in 1965. His name was a blend devised from the names of his sire, QUORUM, and his dam, MARED. It has been pointed out that RUMMY, as he was nicknamed, has a name that if read backwards spells "murder!"

REX, a black Morgan stallion, was the first horse to star independently in western movies, beginning with *Black Cyclone* (1927). His name, Latin for "king," refers to his nickname as "King of the Wild Horses."

REYNOLDSTOWN, the Irish black gelding who won the Grand National in two successive years, 1935 and 1936, was foaled in *Reynoldstown*, Co. Dublin, in 1927 and named for his birthplace.

RIENZI, a black Morgan gelding, was the horse on which the American Civil War commander, Major General Philip Sheridan, made his famous ride from Winchester, Virginia, in 1864 to rally his troops at Cedar Creek, some 14 miles away. The horse was named for the town of *Rienzi*, Mississippi, an earlier objective of a raid.

RIN TIN TIN was one of the most famous "dog stars" of the movies. A German shepherd, he was one of a litter rescued by American airmen from an abandoned German dugout in France in September 1918, and taken back to California. Corporal Lee Duncan named the pup for a small doll that French soldiers carried for luck. The doll's name in turn is essentially meaningless, but if translated would be something like "Tinkerbell." RINTY was originally trained for dog shows, but began his film career in *Where the North Begins* (1923) and from then on never looked back.

RODRIGO DE TRIANO was the horse that won the Two Thousand Guineas at Newmarket in 1992, giving veteran jockey Lester Piggott his 30th British Classic victory. The horse was named for Rodrigo de Triana (*sic*), the sailor who accompanied Christopher Columbus on his first transatlantic voyage in the *Santa Maria* in 1492 and who is said to have been the first to sight land. The name was appropriate for a winning horse in the quincentenary year of Columbus's discovery of America.

ROSABELLE was the favorite palfrey of Mary, Queen of Scots (1542–1587). Her name appears to be of Latin or Italian origin, as a combination of *rosa* and *bella*, and so to mean "beautiful rose." On the other hand, the first part of the name may aptly represent Germanic *hros*, "horse," so that the overall meaning is simply "beautiful horse."

RUMPEL was a cat owned by the English poet Robert Southey (1774–1843). As is often the case with cats, he had a mock "formal" name: "The Most Noble the Archduke Rumpelstilzchen, Marquis Macbum, Earle Tomlemange, Baron Raticide, Waowhler, and Skaratch." See also HURLYBURLYBUSS, above.

RYBOLOV was a tabby cat owned by the Russian composer Alexander Borodin (1833–1887). His name means "Angler," referring to his habit of catching small fish with his paw through ice-holes in the winter. Another cat owned by the musician was DLINENKY ("Longy" or, in modern terms, "Slimline").

SARDAR was the bay gelding presented to Jacqueline Kennedy by President Mohammed Ayub Khan during her visit to Pakistan in 1962. The horse's name is Hindustani for "chief," "headman."

SAVOY was the favorite black horse of Charles VIII of France, named for its donor, the Duke of *Savoy*.

SELIMA was the English writer Horace Walpole's cat, whose demise inspired Thomas Gray's *Ode on the Death of a Favourite Cat Drowned in a Tub of Gold Fishes* (1748). Before writing, Gray checked with Walpole that he had the name right: "Selima, was it? or Fatima." The name is of uncertain origin, but Gray's poem shows that it was stressed on the first syllable (like "Pamela"). The name was perhaps based on Arabic *sālim*, "safe," "secure." A similar name was that of the cat SELMA, owned in his childhood by the English travel writer Augustus Hare (1834–1903). (A sadistic aunt ordered the cat to be hung, as a result of which, it is said, Hare developed a pathological hatred of women and never married in consequence.)

SHASTA was a liger (hybrid offspring of a *li*on father and a ti*ger* mother) born at the Hogle Zoological Garden, Salt Lake City, Utah, in 1948. Her name is said to have evolved from people saying of her as a cub, "She hasta have this, she hasta have that," although *Shasta* is also the name of a North American Indian people of northern California and southern Oregon, and of a famous mountain in northern California said to have mystical attributes.

SHERGAR, the English Derby winner, gained notoriety when he was kidnapped in 1983, never to be seen again. He was foaled in 1978 and has a name that was devised from those of his sire and dam, respectively GREAT NEPHEW and SHARMEEN. *Sher,* moreover, is the Persian word for "lion" and a Muslim name in its own right. SHERGAR was bred by the Aga Khan.

SIMBA was the black and white fox terrier owned by the English naturalist and explorer Cherry Kearton (1871–1940). He originally named her PIP, a "silly, commonplace name" for what seemed to be a "commonplace" dog. She accompanied him to Central Africa, however, and there proved herself to be both fearless and trustworthy. He therefore renamed her SIMBA, the Swahili word for "lion." The name itself was conferred on her by an African chief as a result of her braveness in danger. The name is now generally popular for German shepherds.

SUE hit the headlines in the early 1990s as the dinosaur whose fossil remains were discovered in Indian reservation lands in South Dakota. The remains were claimed by the Black Hills Institute of Geological Research, Hill City, South Dakota, but were counterclaimed by tribal leaders from the reservation, who contended that the dinosaur belonged to them. The justice department eventually decided that SUE was federal property, and in 1992 "she" was duly seized from the Institute by the FBI as "criminal evidence." The dinosaur was named for the Institute researcher who had discovered the remains two years earlier.

TAO is the Siamese cat who accompanied BODGER the bull terrier and LUATH the retriever in the trek across Canada described in Sheila Burnford's *The Incredible Journey* (1961). His name, appropriately, is the Chinese for "way," "path."

TIDDLES was the name of a cat that gained media attention for making his home in the ladies' toilets at Paddington Station, London. He lived there 13 years, dying in 1983. His name was originally the conventional one (see Chapter 2, page 14), but later took on added senses: first, callers regularly fed him, until he became grossly overweight, or anything but "tiddly"; second, *tiddle* is a colloquial word meaning "urinate."

TIGER has been the name of various well-known cats, from the pet of the English author Charlotte Brontë (1816–1855) to the gray-striped alley cat adopted in the 1920s by American president Calvin Coolidge (1872–1933).

TIMBER DOODLE was the small white spaniel given to Charles Dickens on his visit to New York in 1842. His name is a term for the American woodcock. On his return to England, Dickens wrote: "I have changed his name to Snittle Timbery as more sonorous and expressive." He took the name from that of a character in one of his own novels, *Snittle Timberry* (*sic*), a performer in Vincent Crummles' touring company in *Nicholas Nickleby* (1839).

TOM QUARTZ was a kitten owned by American president Theodore Roosevelt. He was named for Dick Baker's cat in Mark Twain's *Roughing It* (1872), himself so called for his master's quartz mining.

TONY was famous as American movie actor Tom Mix's "Wonder Horse" in the 1920s. His name is the American slang word for "classy" (i.e., having "tone") rather than the short form of *Anthony*.

TOPSY was the donkey owned by the Scottish Episcopal bishop, the Rt. Rev. George Sessford (born 1928). He explains how the animal came by her name, which in full was TOPSY O'CEDAR:

> By common consent we named the new foal Topsy. It was a highly suitable name in every way—until we discovered that the wife of a local doctor was also called Topsy, and as she, too, was a member of the Episcopal church, we had a few embarrassing moments. [...] We added the name "O'Cedar" for two reasons. First, Topsy was of Irish descent and so "O" rather than "Mac" seemed a suitable prefix. Second, we discovered that "O'Cedar" liquid polish worked wonders on her tiny hoofs as we prepared her to take part in the nativity play [from "A Highland

Bishop and His Donkey," in Bill Annett and Marta Annett, *For the Love of Animals;* see Bibliography].

As a general name, TOPSY suggests both "top" and smallness, with specific reference to the little black slave girl in *Uncle Tom's Cabin*. It is thus fairly common for a young or small animal, such as the Pekinese owned by the British novelist Dennis Mackail (1892–1971). Mackail included the dog's name in the title of his autobiography, *Life with Topsy* (1942), which he additionally dedicated to her.

TRAVELLER was the gray horse bought by Confederate General Robert E. Lee in 1861 during the Civil War. His original name was JEFF DAVIS, for the then American senator and future president of the Confederate States of America, *Jeff*erson *Davis*. Lee changed his name, however, because the horse *travelled* so well.

TREPP was billed as "the world's top police dog" in 1978, after he had uncovered many illegally imported drugs as an American sniffer dog. A golden retriever, his name was short for the formal name of IN-TREPID.

TRIGGER was famous as Roy Rogers' golden palomino, the star of many musical westerns. His name is fitting for a horse whose master is a gunslinger.

TUNIS was the black gelding bought by General Georges Boulanger, French minister of war, for his first Bastille Day review of troops in 1886. The horse was named for *Tunis,* capital of the French protectorate of Tunisia, north Africa, where Boulanger had commanded the army of occupation.

TYRAS was the name of two of the several black great Danes owned by the famous Prussian statesman Prince Otto von Bismarck. *Tyras* is the ancient name of the Dniester River, now in Ukraine.

VENUS was the black cat owned by the Welsh-born poet and author W.H. Davies (1871–1940), and so named for her beauty, intelligence, and love. Davies had a dog with a similarly sensitive name, BEAUTY BOY.

VONOLEL, the Arabian horse who was the favorite charger of Field Marshal Lord Roberts in the Second Afghan War and later in India, was bought by his master in 1875. He was named for a warrior of the Lushai people, on the India-Burma border, against whom Roberts had led an expedition in 1872.

VULCAN was the large dog owned by the first American president, George Washington (1732–1799). His name is a "strong" one, that of the god of fire and metalworking in Roman mythology. Other presidential dogs, with names typical of their day, were CAPTAIN, FORESTER, MOPSEY,

ROVER, SEARCHER, SWEET LIPS, TASTER and TIPPLER. Washington's successors in the White House were not noted dog lovers until the next century, and even Abraham Lincoln (1809–1865) was associated with only one dog, the mongrel FIDO, owned by his sons.

WASHOE, the first chimpanzee to communicate in words (in the 1960s), was named for the Nevada county that was the home of R. Allen and Beatrice Gardner, the behavioral psychologists who taught her.

WESSEX was the wirehaired terrier of the English author Thomas Hardy (1840–1928), whose poems and novels were largely set in the ancient kingdom of this name in southern England.

WILBERFORCE was the official "cat in residence" at No. 10 Downing Street, London. He arrived in 1973 and served under four prime ministers (Edward Heath, Harold Wilson, James Callaghan, Margaret Thatcher) until his death in 1988. He was named for the British politician William *Wilberforce* (1759–1833), who secured the abolition of the slave trade. His own aim was to secure the abolition of mice in No. 10. He was succeeded as prime-ministerial mouser by HUMPHREY.

WINSTON was well known as the London police horse who became the royal charger ridden by Queen Elizabeth II at the ceremony of Trooping the Color in the 1950s. He was named partly for the British prime minister in World War II, *Winston* Churchill, but also partly because he joined the Metropolitan Police Force in 1944, a year when the "Met" named their horses with the identifying letter W.

YUKI was the small white dog adopted by American president Lyndon B. Johnson in 1966 after the loss of his beagle HIM (see above). His name is Japanese for "snow."

# 11

# *Animals of Myth, Legend, and Cartoon*

Some of the best-known animals of myth and legend are those in the Greek and Roman classics, in particular dogs and horses.

ARGUS, whose name represents Greek *argos,* "shining," "bright," was the faithful dog of Odysseus (otherwise Ulysses), who recognized his master on the latter's return home after a 20-year absence and died from joy. The Greek adjective was often used of dogs and was interpreted in the sense "swift-footed," because the rapid motion of a dog's legs as it walks or runs was seen to be accompanied by a kind of flickering light.

Then there was the famous three-headed (or 50-headed) CERBERUS, watchdog of the Underworld. His name is probably not Greek at all, but was popularly understood to represent Greek *Ker berethrou,* "evil of the pit," from *Ker,* the name of the evil spirit of death, and *berethron,* an Ionic form of *barathron,* "gulf," "pit." Icarius (not to be confused with Icarus) had a dog named MAERA, said to be from Greek *marmairō,* "I sparkle," "I glisten," for the same reason that gave the name of ARGUS (above). Orion, the handsome hunter, had two hunting dogs, ARCTOPHONOS, "bear-killer" and PTOOPHAGOS, "glutton of Ptoüs," the latter being a young man about whom little is known. Procris, daughter of Erechtheus, had a dog named LAELAPS, given her by the goddess Diana when she fled her husband Cephalus. The dog must have been a "fast and furious" beast, for its name represents Greek *lailaps,* "hurricane," "whirlwind."

Diana (or Artemis) is famous for changing Actaeon into a stag when he had seen her bathing, with the fearful consequence that he was torn to pieces by his own hounds, all 50 of them. Their names are preserved in legend, and are given in Part 1 of Appendix III.

Helios, the sun god, had a fine team of steeds: ACTAEON, "brightness," AETHON, "fiery red," AMETHEA, "non-loitering," BRONTE, "thunderer," ERYTHREOS, "red-producing," LAMPOS, "bright," "lamplike," PHLEGON, "blazing," and PYROCIS, "burning." The horses of Aurora (or Eos), goddess of the dawn, were ABRAXAS, of uncertain origin, EOOS, "dawn," and PHAETON, "shining," while the horses of Pluto, god of the Underworld, were ABASTER, "away from the stars," ABATOS, "inaccessible," AETON, "eagle-swift," and NONIOS, perhaps a corruption of Greek *nōnymos,* "nameless."

The horse of Hercules, formerly belonging to Neptune, was ARION, "martial." BALIOS, "swift," was one of the horses given by Neptune to Peleus, and later belonging to Achilles. Like XANTHUS, "reddish-yellow," its sire was Zephyrus, the west wind. A horse swifter than the wind was CERUS, "fit," owned by Adrastus, king of Argos. CYLLAROS, named for *Cylla* in Troas, was a horse owned by the twins Castor and Pollux, as was HARPAGUS, "snatcher," "one who carries off." DINOS, "marvel," was owned by Hector, as was GALATHE, "milk-colored." HIPPOCAMPUS was one of Neptune's horses who had only two legs, its hindquarters being those of a fish. Its name derives from Greek *hippos,* "horse," and *kampos,* "sea monster."

PEGASUS, one of the most familiar horses of classical mythology, has his own entry in Chapter 12, page 156.

In Egyptian mythology, APIS was the sacred bull worshipped at Memphis. His name is a classical Greek form of Egyptian *Hapi,* itself probably meaning "the hidden one." The bull was regarded as the incarnation of the widely worshipped god Osiris, and his name subsequently combined with that of the god to give *Serapis,* a new Egyptian god also venerated at Memphis.

The names of animals in Muslim mythology are of later appearance. BORAK or AL BORAK was the white animal, in essence a winged horse, that carried Mohammed to the seven heavens. Its name means "lightning" (modern Arabic *bark*). Mohammed's white mule was FADDA, "silver" (modern Arabic *fadda*).

The names of several horses are well-known from Norse legends. They include ARVAK, "early waker," who draws the sun's chariot, ALSVID, "all swift," who draws the chariot of the moon, HRIMFAXI, "frost mane," the horse with the mane of rime that Nott, goddess of night, drove across the sky, and SKINFAXI, "shining mane," driven across the heavens by Dag, god of day, to brighten the earth and sky with the light from his mane. There was also SLEIPNIR, "slipping one," the eight-footed steed of Odin, who could run faster than the wind. His sire was SVADILFARI, "track traveler," and his dam was the god Loki, who had turned into a mare. He himself sired GRANI, "gray."

A famous character of Celtic mythology was the legendary hero Cuchulain. His name represents Irish *cú Chulain,* "Culan's hound." He was so called as he had offered to take the place of the fearful watchdog of the smith Culan, which he had killed. In the account of his adventures told by James Macpherson in *Fingal* (1762), in which he appears as Cuthullin, he has his own hound LUATH, from the identical Irish word meaning "swift," "nimble." The name has been adopted for several real and fictional dogs in modern times. (See LUATH in Chapter 10, page 122.)

Fingal (known as Finn mac Cool in Ireland) himself had a hound named BRAN, literally "raven," otherwise "chief." This name has also been adopted for more recent dogs, again both real and fictional. An example of the latter is the bitch ("a four-legged, brindled, lop-eared, toad-mouthed thing") owned by Raphael Aben Ezra in Charles Kingsley's novel *Hypatia* (1851).

There are no named animals in the Bible, although the Old Testament has Balaam's talking donkey (Numbers 22.28), the New Testament has Jesus riding into Jerusalem on a donkey (Luke 19.35), and the Apocrypha has Tobias and the angel Gabriel setting out for a journey accompanied by Tobias's dog (Tobit 5.16). However, see TOBY in Chapter 12, page 162.

A few named horses and dogs appear in the works of Shakespeare, with CRAB actually coming on stage. Individual horses such as ROSINANTE and BRIGADORE are also famous from contemporary literature. The dog TRAY, too, is familiar in literature of various kinds from Shakespeare onward. (See these and others in Chapter 12.)

In a category of their own are the curious names recorded mainly in the 17th century for "familiars." A "familiar," also sometimes known as an "imp," was an evil spirit in animal form who was said to attend a witch. The form that the spirit took varied, but was popularly believed to be that of a cat or frog. Matthew Hopkins (died 1647) set up as a witch hunter, and in his famous book *The Discovery of Witches* (1647) names several of the "familiars." Thus one Elizabeth Clark, an old, one-legged beggar woman, gave the names of her "imps" as HOLT ("a white kitling"), JARMARA (a "fat spaniel" without legs), SACKE AND SUGAR (a "blacke rabbet"), NEWES (a "polcat"), and VINEGAR TOM (a greyhound with the horned head of an ox). Another witch had ILEMAUZAR or ELEMAUZER, PYEWACKETT, PECKE IN THE CROWNE, and GRIEZZELL GREEDIGUTT, names, comments Hopkins, "which no mortal could invent." (For a modern literary revival of one of these, see PYEWACKET in Chapter 12, page 157.) A frontispiece to Hopkins' book, in the form of a crude woodcut, shows Hopkins himself as "Witch Finder Generall," Elizabeth Clark with her "imps," and the other, unnamed witch giving the names of her own "imps." They do not appear in the picture but are presumed to be cats.

Earlier, in 1617, a cat named RUTTERKIN featured prominently in the trials of the so called "Belvoir Witches," the three members of the Flower family (mother Joan and daughters Margaret and Phillipa) who worked at Belvoir Castle, Leicestershire, and who were accused of bewitching the children of the 6th Earl of Rutland.

The precise meaning or significance of many of these names is obscure. No doubt there was originally some occult reference.

In *Old Possum's Book of Practical Cats* (1939), T.S. Eliot quotes names that in many cases are amusing versions of the names of "familiars" like these. In the opening poem, "The Naming of Cats," for example, he says that "a cat needs a name that's particular," and by way of example quotes MUNKUSTRAP, CORICOPAT, BOMBALURINA, and JELLYLORUM, among others. Some of the cat characters in the book also have names of this type, such as MUNGOJERRIE and RUMPELTEAZER, while MR. MISTOFFELEES is obviously a whimsical variant on the name of *Mephistopheles* himself, as the devil in medieval mythology to whom Faust sold his soul. In this same first poem Eliot makes his famous statement about every cat having three names, the third being one "that no human research can discover," and that the cat alone knows as his "deep and inscrutable singular Name." It is this name, according to Eliot, that the cat is contemplating when he is "in profound meditation." These whimsically mystical observations, incidentally, are largely absent from Andrew Lloyd Webber's popular musical *Cats* (1981), based on the book. (For one such stage cat, see GRIZABELLA in Chapter 12, page 150.)

The names of animal characters have been popularized in a more orthodox manner by other modern writers, such as Walter Scott, Charles Dickens, R.S. Surtees, and Beatrix Potter (see all these in Appendix III), while individual children's books have made famous the names of such as the horse BLACK BEAUTY and the dog CRUSOE.

Many animals appear in fiction under their generic name, which is raised to the status of a proper name for fictional purposes. Good examples are RAT, MOLE, BADGER and MR. TOAD in Kenneth Graham's still popular children's book, *The Wind in the Willows* (1908). Such names are similarly found in the children's books by A.A. Milne, where PIGLET, RABBIT, OWL and others appear in "plain" form, with KANGA, ROO and TIGGER only thinly disguised. Where the names coincide exactly or very nearly with the generic words for the animal like this, their meanings are obvious. They will therefore not appear again in Chapter 12. But names such as EEYORE and WINNIE-THE-POOH duly have their entries.

In more recent times still, many animals have become extremely well known through their realization as cartoon characters. Although they are

## 11. Animals of Myth, Legend, and Cartoon

mostly familiar from comics or comic strips, from animated cartoons in movie theaters and on television, or from radio programs, and are thus not properly "fictional" in the usual sense of being characters in novels and short stories, they have a rightful place in the popular pantheon of animal names. Names of puppet animals should also not be overlooked, since the characters who bear them are in a way more "real" than those appearing in cartoons. Many such names are of American origin, such as the famous cartoon creations of Walt Disney and the Hanna-Barbera team. A brief survey of the characters and their names, which equally include those of puppets, is therefore appropriate at this point. (The survey does not generally consider the many characters who first appeared in book form before transferring to an animated cartoon interpretation, such as BABAR, BAMBI, DUMBO, or JIMINY CRICKET.)

The name of a cartoon or puppet character frequently comprises a personal name followed by the generic name for the particular animal, the latter serving as a "surname." The classic example is MICKEY MOUSE, who can be otherwise understood to be a mouse named "Mickey." The particular personal name adopted for the character is usually an arbitrary one that either rhymes with the animal's generic name or alliterates with it (or both). Typical examples (apart from MICKEY MOUSE himself) are ANDY PANDA, BARNEY BEAR, CLAUDE CAT, DONALD DUCK, ROGER RABBIT, and ROLAND RAT. However, the first name is not restricted to the conventions that normally apply for personal names, so can be a descriptive nickname or simply a standard word of some kind, often with a diminutive ending such as -Y to suggest a genuine first name. Examples are BUGS BUNNY, DAFFY DUCK, DINKY DUCK, GANDY GOOSE, KRAZY KAT, MIGHTY MOUSE, PORKY PIG, and WOODY WOODPECKER. DEPUTY DAWG may be added here, too, with a title as first name, while HUCKLEBERRY HOUND is something of a blend of both types. Sometimes the second word of the name may be a distinctive feature of the animal's person, giving such names as BASIL BRUSH (a fox) and TEDDY TAIL (a mouse). In other instances, the generic name can come first, as for TIGER TIM.

Another popular formula, however, is "X the Y," where "X" is a personal name (or something like it) and "Y" is the generic name for the animal. Typical examples here are ALBERT THE ALLIGATOR, DENNIS THE DACHSHUND, FELIX THE CAT, FLIP THE FROG, HUGO THE HIPPO, KORKY THE CAT, LARRY THE LAMB, LENNY THE LION, LIPPY THE LION, MUFFIN THE MULE, and PETE THE PUP. Many of these are popularly known just by their main name, so that they are simply DENNIS, FELIX, FLIP, HUGO, KORKY, and so on. (LARRY THE LAMB and DENNIS THE DACHSHUND appeared in a series of nursery story books by the British artist and toymaker S.G. Hulme

Beaman from the mid-1920s. In real life the former name has become associated with the noted English newspaper editor Sir Larry Lamb, born in 1929, the year when the stories were first broadcast on the BBC's popular *Children's Hour* program.) The skunk PEPE LE PEW is named for his distinctive aromatic attribute, so has a name similar to that of TEDDY TAIL.

Many such names alliterate. Some characters have names that do not, however, such as FOZZIE BEAR, FRED BASSET, BOSS CAT, SCOOBY-DOO, TOP CAT, YOGI BEAR, or (in the "X the Y" form) ALEXANDER THE MOUSE, FRANCIS THE MULE, HOWARD THE DUCK, and KERMIT THE FROG. MISS PIGGY also has a nonalliterative name, the first word of which is properly a title.

Paired names may alliterate (or rhyme) or not. Typical examples are CHIP an' DALE (chipmunks), HECKLE and JECKLE (magpies), PINKY and PERKY (pigs), ROOBARB and CUSTARD (dog and cat), RUFF and REDDY (dog and cat), SPIKE and TYKE (dogs), TINGHA and TUCKA (koalas), and the famous TOM and JERRY (cat and mouse).

Names like these are all in the mold of their human cartoon equivalents, such as *Dennis the Menace, Desperate Dan,* and *Mutt and Jeff.*

Names of associated cartoon characters, whether as "family" or "friends," are frequently of mixed type. Examples of such names familiar to British children from 1920, when the comic-strip characters first appeared in the *Daily Express,* are those of RUPERT BEAR and his pals ALGY PUG, BILL BADGER, BINGO THE BRAINY PUP, and EDWARD TRUNK the elephant.

Some cartoon or puppet characters have names consisting of a single word, whether this is a regular personal name or, more commonly, a nickname or distinctive word. Well-known examples are the dog BONZO, the moose BULLWINKLE, the cat GARFIELD, the dog GOOFY (originally planned to be DIPPY DAWG, however), the dog PLUTO, the opossum POGO, and the dog SNOOPY. The cat SYLVESTER and the canary TWEETY PIE made separate film debuts before teaming as a pair. The latter's name puns on both *tweet* and *sweetie pie* as a term of endearment. Names of this "one-word" type are sometimes formed by running two words together, as for BAGPUSS the cat or SNAGGLEPUSS the lion, among others.

Some names of puppets are inexorably bound up with those of their creators or operators, especially where the operators appear with the puppet in a dual role. Even so, in themselves they are genuine animal names, and so likewise deserve their place here. Among the most familiar to have appeared on television alone, and by no means with just children as their audience, are Keith Harris and CUDDLES and ORVILLE (both somewhat indeterminate specieswise). Rod Hull and EMU, Shari Lewis and LAMB

*11. Animals of Myth, Legend, and Cartoon* 139

CHOP, Roger de Courcey and NOOKIE BEAR, Harry Corbett and SOOTY (the bear), and Bob Carolgees and SPIT (the "punk dog").

With the exception of the names of classical mythology, whose meanings were considered above, many of the fictional names in this chapter, including those of well-known cartoon characters, have their individual entries in Chapter 12, following.

# 12

# *Animals in Fiction*

From our earliest story books, we discover that animals are named very much as they are in real life. In Enid Blyton's *Shadow the Sheepdog* (1969), for instance, the young reader is not in the least surprised to find that the boy Johnny calls his dog SHADOW because "he follows me about just like my shadow does."

The following is a selection of some of the hundreds of names of animals in fiction, including familiar characters from movies, cartoons, radio and television programs, together with their origins, where these are known or can reasonably be deduced. The selection is chronologically wide-ranging, taking in individual animals from medieval romances as well as from 20th-century comic strips and commercials. Since many of the characters from children's books or animated cartoons are known internationally, examples of their names in non–English languages are provided for interest.

AKELA is the "great gray Lone Wolf, who led all the Pack by strength and cunning" in Rudyard Kipling's *Jungle Book*s (1894, 1895). The stories are set in India, so that his name, together with those of most of the other animals, is Hindi in origin. As "Lone Wolf" his own name thus represents Hindustani *akela,* "lonely," "solitary."

ALIDORO is the bull mastiff in Carlo Collodi's children's classic *Pinocchio* (1883). The dog cannot swim, so he is a "dry" dog, with a name from Italian *alidore,* "dryness."

APOLLYON is Herman von Arnheim's jet black horse in Walter Scott's *Anne of Geierstein* (1829). According to the novel, the fact that he bore the name of the biblical "angel of the bottomless pit" (Revelation 9.11) "was secretly considered as tending to sanction the evil reports which touched the house of Arnheim."

ARUNDEL is the horse of the eponymous hero of the medieval verse romance *Bevis of Hampton*. The name represents the Old French word for "swallow" (modern French *hirondelle*), denoting swiftness.

ASLAN is the great lion in the *Narnia* series of children's books (1950–56) by C.S. Lewis. His name is simply the Turkish word for "lion." Lewis is said to have come across it when reading the *Arabian Nights*.

ASTA is the schnauzer bitch in Dashiel Hammett's detective story *The Thin Man* (1934), and also the fox terrier (played by a male named Skippy) in the movie version of this (1934). Her name is Spanish and Italian for "spear," "lance," no doubt referring to her speed of running or attack.

BABAR is the "King of the Elephants" in *The Story of Babar* (1931), written and illustrated for children by Jean de Brunhoff. His name suggests Hindustani *bābā*, which can mean both "father" and "child," and is a term of respect.

BAGHEERA is the black panther in Rudyard Kipling's *Jungle Books* (1894, 1895). His name is based on Hindustani *bāgh*, "tiger," with a second half intended to represent what Kipling described as "a sort of diminutive."

BAIARDO *see* BAYARD

BALOO is the "sleepy brown bear" in Rudyard Kipling's *Jungle Books* (1894, 1895). His name simply represents Hindustani *bhālu*, "bear."

BAMBI is the fawn in the story named for him (1928) by Felix Salten and in the subsequent Disney animated cartoon (1942). His name is a short form of Italian *bambino*, "baby."

BAVIAAN is the wise baboon in Rudyard Kipling's short story "How the Leopard Got His Spots" in *Just So Stories for Little Children* (1902). The name is actually a form of *baboon* itself.

BAVIECA (or BABIECA) is the legendary steed of the famous Spanish hero El Cid ('the lord'). His name represents Spanish *babieca*, "blockhead," "dolt." The story goes that El Cid had chosen a rough colt for his charger, instead of a trained battlehorse. His godfather called him "simpleton" for making such an unsuitable choice, and the Spanish champion transferred the name to his horse. Compare the similarly derogatory name of ROSINANTE, below.

BAYARD is the magic horse in French legend that was given by Charlemagne to the hero Renaud de Montauban, who features as Rinaldo in Boiardo's *Orlando Innamorato* (1487) and in Ariosto's *Orlando Furioso* (1532). His name is French for "bay," describing his coloring. In the two works mentioned his name appears in its Italian form of BAIARDO. The name was common for any bay horse.

BELZEBUB is the colt who is Wilfred Thorne's favorite young steed in Anthony Trollope's novel *Barchester Towers* (1857). His name is a formerly common variant form of *Beelzebub* (Hebrew, "lord of the flies"), alluding to his "devilish" or at any rate capricious nature.

BERGANZA is the talking dog in the tale *El Coloquio de los Perros* ("The

Colloquy of the Dogs") in Cervantes' *Novelas Ejemplares* ("Exemplary Novels") (1613). His name is a form of Spanish *bergante*, "rascal."

BEVIS is Sir Henry Lee's "faithful mastiff or bloodhound" in Walter Scott's *Woodstock* (1826). Scott had already used the name for the hero's horse in his long poem *Marmion* (1808). He doubtless adopted it from that of the hero of the medieval romance *Bevis of Hampton*. The dog was modeled on Scott's beloved MAIDA (see Chapter 10, page 122).

BIGWIG is the brave older rabbit who helps several does escape in Richard Adams' classic animal novel *Watership Down* (1972). His name is explained early in the story: "He had a curious, heavy growth of fur on the crown of his head, which gave him an odd appearance, as though he were wearing a kind of cap. This had given him his [rabbit-speak] name, *Thayli*, which means, literally, 'Furhead,' or as we might say, 'Bigwig'" (Chapter 2).

BINGO is the black and white dog who, with Jack the sailor boy, has appeared on Cracker Jack packs in the United States since 1919. Jack was modeled on Robert Rueckheim, whose grandfather, F.W. Rueckheim, founded a popcorn business in Chicago in 1873. BINGO was originally the name of young Robert's dog. The origin of the name is uncertain. As a standard word, *bingo* was in use from the 17th century as a slang term for brandy, and it is conceivable that the name was adopted for dogs as a sort of familiar equivalent of BRANDY itself, as the name of a brown dog. It was also familiar in the first quarter of the 20th century as the name of the boy with large ears who featured in the comic strip *The Bingville Bugle*. (He himself gave the adopted first name of the popular actor and singer Bing Crosby, who also had prominent ears.) Compare BINGO THE BRAINY PUP as a cartoon character in Chapter 11 (page 138), and see also BINKIE, below.

BINKIE is Gilbert Belling Torpenhow's dog in Rudyard Kipling's novel *The Light That Failed* (1890). The name is perhaps an affectionate form of BINGO (see above). It is otherwise familiar from the nickname adopted by the noted British theatrical manager, Hugh "Binkie" Beaumont (1908–1973). In his case, it is said to have originated from the humorous use of a local Welsh word for a dark-skinned or dirty child given him by some women ("He's a *real* binkie, isn't he!") as he lay one day in Cardiff in his baby carriage, clean, fair-skinned, and golden-haired.

BLACK BEAUTY, with his self-descriptive name, is the hero of the novel by Anna Sewell named for him, subtitled *The Autobiography of a Horse* (1877). Possibly through associations with "Beauty and the Beast" he is sometimes assumed to be a mare, and is mistakenly referred to as such in *The Oxford Companion to English Literature* (5th ed., 1985).

BLANCH is one of three dogs mentioned by the distraught king in Shakespeare's *King Lear* (see TRAY, below). The name means "white," from the French.

BOO-BOO is the little bear who is the companion of YOGI BEAR (see below) in the American Hanna-Barbera television cartoons (1958-62). His name may be arbitrary but is not uncommon for a *baby* or human-seeming animal. The London *Daily Mail* of February 17, 1935, reported that BOOBOO, a female chimpanzee in London Zoo, had given birth to a baby called JUBILEE (for the 25th anniversary of the accession of George V). See the name also in Appendix IV.

BOOTS is the black Aberdeen terrier, "son of Kildonan Brogue—Champion Reserve—V.H.C.—very fine dog," who is the narrator in Rudyard Kipling's *Thy Servant a Dog* (1930), so named for his close association with his master's shoes and as a pet form of his pedigree name. His companion is SLIPPERS, with a similar name, and the same association with his mistress's footwear.

BOSS CAT *see* TOP CAT

BOXER is John Peerybingle's lively mongrel in Charles Dickens' *The Cricket on the Hearth* (1846), in which he is named for his pugnacity, not his breed. The name is also that of the carthorse, "an enormous beast, nearly eighteen hands high, and as strong as any two ordinary horses put together," in George Orwell's novel *Animal Farm* (1945). He was doubtless named for his power and strength, like that of a *boxer*.

BRER BEAR, BRER FOX, BRER RABBIT, BRER TARRYPIN (a *terrapin* or turtle) and BRER WOLF are characters in the *Uncle Remus* stories (1880) by Joel Chandler Harris. *Brer* means "brother."

BRIGADORE is the "loftie steed" of Sir Guyon in Edmund Spenser's *The Faerie Queene* (1590). His name, meaning "golden bridle," was adopted (and adapted) from that of BRIGLIADORO (see below).

BRIGLIADORO is the famous charger of Orlando (Roland) in Ariosto's poem *Orlando Furioso* (1532), in which he is second only to Rinaldo's magic horse BAIARDO (see BAYARD). His name means "golden bridle," from Italian *briglia d'oro,* "bridle of gold." The horse BRIGADORE (see above) was named after him.

BUCK is the dog, part St. Bernard and part Scotch shepherd, who is the central character of Jack London's novel *The Call of the Wild* (1903). His name suggests masculinity and strength. A sequel to the novel was *White Fang* (1906) (see WHITE FANG, below).

BUGLE ANN is the foxhound with the loud bay who is the central character of MacKinlay Kantor's popular story *The Voice of Bugle Ann* (1935), set in Missouri in the 1930s. The hound's name alludes to her powerful voice. *Bugle* (or *bugle-horn*) is properly a term for a hunting horn, so called because it was originally made from the horn of a *bugle*, or wild ox.

BULL'S-EYE is Bill Sikes' "white shaggy dog, with his face scratched and torn in twenty different places" in Charles Dickens' *Oliver Twist* (1839). His name suggests that he was a pugnacious dog, like one used for bull-baiting. He does not appear to have actually been a bulldog or bull terrier, however. It is presumably a coincidence that his name resembles that of his master.

CADPIG is the smallest and prettiest of the 15 puppies of PONGO (see below) in Dodie Smith's *The Hundred and One Dalmatians* (1956). As the story explains: "When pigs have families, the smallest, weakest piglet is often called the cadpig. Mr. Dearly always called the tiny puppy 'Cadpig,' which can be a nice little name when spoken with love."

CAPILET is Sir Andrew Aguecheek's horse in Shakespeare's *Twelfth Night* (1599). The name is a diminutive of *caple,* an old word meaning simply "horse," itself directly related to Latin *caballus,* "hack," "nag," and so to modern French *cheval,* "horse." Compare CAVALL, below.

CAPTAIN FLINT is the pirate chief Long John Silver's parrot in R.L. Stevenson's adventure romance *Treasure Island* (1883). The parrot is named for the captain whose buried treasure is the subject of the story.

CAT is the cat in the novel by Truman Capote, *Breakfast at Tiffany's* (1958). A real cat played his part in the movie version (1961).

CAVALL is "King Arthur's hound of deepest mouth" in Tennyson's *Idylls of the King* (1859). His name actually means "horse," from a word related to modern Welsh *ceffyl* or Irish *capall.* This implies that in some way he was "horselike," perhaps having a large head or sturdy body. Compare the name of CAPILET, above.

CHARLOTTE is the name of the central character, a large gray spider, in E.B. White's children's novel *Charlotte's Web* (1952). The spider introduces herself to her new friend, the little pig WILBUR:

> "What is your name, please? May I have your name?"
> "My name," said the spider, "is Charlotte."
> "Charlotte what?" asked Wilbur, eagerly.
> "Charlotte A. Cavatica. But just call me Charlotte."

CHEE CHEE is the monkey who is one of the companions of Doctor Doolittle in the stories by the English children's writer Hugh Lofting (1886–1947). Other animals to appear are GUB GUB the pig and POLYNESIA the parrot. CHEE CHEE's name is onomatopoeic, representing the high-pitched chattering of a monkey. The pig's name is similar, suggesting rooting or gobbling sounds. The parrot's name is of course a fanciful elaboration of POLLY. The first book in the series was *The Story of Doctor Doolittle* (1920), published the year after Lofting settled in the United States.

CHIL is the kite in Rudyard Kipling's *Jungle Book*s (1894, 1895). His name simply represents Hindustani *chīl*, "kite," no doubt itself coming from the bird's thin, shrill cry. (The English word *kite* probably has a similar onomatopoeic origin. It has evolved from Old English *cȳta*, with a long, thin vowel sound like that in modern French *rue*.)

CHIP AN' DALE are the pair of chipmunks who made their debut in the Disney cartoon *Private Pluto* (1943). Their names are obviously based on *Chippendale*, itself doubtless associated with *chipmunk*. In French they are known as *Tic et Tac*, in Italian as *Cip e Ciop*, in Dutch as *Knabbel en Babbel*, but in German simply as *A- und B-Hörnchen* ("Chipmunk A and B").

CLAVILENO is the magic steed ridden by Don Quixote in the satirical romance by Cervantes of 1605 that bears his name. The horse is made of wood, and has a peg for guiding it. Its name describes this distinctive feature, and so means "wooden peg," from Spanish *clavija*, "peg," and *leño*, "wood." Don Quixote also had the famous horse ROSINANTE (see below).

CLOVER is one of the two carthorses (the other is BOXER) in George Orwell's satirical novel *Animal Farm* (1945). She is named for the flowering plant that is grown on farms as fodder for cattle.

COLONEL, the English sheepdog who organizes the puppies' escape in Dodie Smith's *The Hundred and One Dalmatians* (1956), is so named as he is "a perfect master of strategy."

CRAB is the only dog who appears on stage in a Shakespeare play. He appears as Launce's dog in *The Two Gentlemen of Verona* (1594), and is described by his master as "the sourest-natured dog that lives." Hence his name, referring to his *crabby* character.

CRUSOE is the dog hero of R.M. Ballantyne's adventure novel *The Dog Crusoe* (1861). He is a Newfoundland puppy rescued by a young North American hunter and obviously named for the famous fictional castaway Robinson *Crusoe*. However, the story itself tells how an illiterate fisherman, who had not heard of Robinson Crusoe, "had got him from a friend, who had got him from another friend, whose cousin had received him as a marriage-gift from a friend of *his*," and "each had said to the other that the dog's name was 'Crusoe,' without reasons being asked or given on either side." Crusoe's great admirer and "little friend" in the novel is the dog GRUMPS.

CURTAL is a horse owned by Lord Lafeu in Shakespeare's *All's Well That*

*Ends Well* (1602). His name represents the standard word for a horse with its tail cut short (hence modern *curtail*).

CUT is the horse that the First Carrier asks the ostler to look after at the Rochester inn in Shakespeare's *Henry IV, Part I* (1597): "I prithee, Tom, beat Cut's saddle, [...] the poor jade is wrung in the withers." The name may describe a horse with its tail cut (like CURTAL, above) or relate to a gelding, who has been *cut*, i.e., castrated.

DAPH and JUNO are Mr. Wardle's gundogs in Charles Dickens' *The Pickwick Papers* (1837). They have arbitrary classical names, DAPH being short for *Daphne*.

DIAMOND is the fictional dog said to have belonged to Sir Isaac Newton. A story tells how in 1693 Newton returned to his college rooms in Cambridge to find that his precious scientific notes had been burned by a candle that the dog had knocked over. But Newton never owned a dog of this name. The name is fairly common for an animal that is a "jewel."

DIOGENES is Dr. Blimber's dog, bought by Toots as a present for Florence Dombey in Charles Dickens' *Dombey and Son* (1847). Not only did he live in a tub, as the famous Greek Cynic philosopher Diogenes is said to have done, but *Cynic* is itself a word deriving from Greek *kunikos*, "doglike." His name thus has a double punning allusion. His pet name is DI.

DOBBIN is Old Gobbo's horse in Shakespeare's *The Merchant of Venice* (1596): "Thou hast got more hair on thy chin than Dobbin my thill-horse has on his tail." As explained in Chapter 2 (pages 17–18) the name is a generic one for a farm horse or old nag. (A *thill-horse* is a draft horse.)

DONALD DUCK, the famous cartoon character, made his first appearance in the Disney cartoon *The Wise Little Hen* (1934). His sweetheart, DAISY DUCK, joined him three years later in *Don Donald* (1937). Both names are fairly arbitrary alliterations on *duck*, so are similar to those of MICKEY MOUSE and MINNIE MOUSE. DONALD keeps his name in many non-English languages, although in Italian he is known simply as *Paperino*, "Little Goose," while DAISY is *Paperina*, "Little Gander."

DOUGAL is the long-haired dog who was the central character in the cult television puppet series *The Magic Roundabout*, broadcast by the BBC in the latter half of the 1960s. The dog's name was doubtless based on *dog* itself, and a similar origin produced the names of his companions ZEBEDEE, the *bee*, and BRIAN, the *snail*. On the other hand, ERMINTRUDE, the cow, was given a mock-solemn name, smacking of Victorian romantic fiction, while DYLAN, the guitar-playing rabbit, was surely so named as a compliment to the rock singer and guitarist Bob *Dylan*, by then a major celebrity.

DUMBO is the baby elephant with big ears in the Disney cartoon (1941) named for him. His name suggests a combination of *dumb*, for his initial slowness, and JUMBO, a famous elephant name (see Chapter 10, pages 120–21).

EEYORE is the old gray donkey in A.A. Milne's children's classic *Winnie-the-Pooh* (1926). His name is imitative of a donkey's cry or "hee-haw."

EMPRESS OF BLANDINGS is the black Berkshire sow owned by Lord Emsworth in the novels by P.G. Wodehouse. She is named for Blandings Castle, Lord Emsworth's county seat, and first appears in the novel *Pig-Hoo-o-o-ey!* (1927).

FELIX is the famous black cartoon cat, first appearing in *Feline Follies* (1929). The cat was so named to suggest "felicity" (Latin *felix*, "happy"), as distinct from the bad luck that black cats are generally said to bring. His name also suggests the scientific name for the cat genus, *Felis*, itself from Latin *feles*, "cat." A truly felicitous name, therefore!

FERDINAND is the bull in the famous picture book *The Story of Ferdinand* (1936) by the American writer and illustrator Munro Leaf. He is a young Spanish bull who likes "to sit just quietly and smell the flowers," and was given a typically Spanish name (originally *Ferdinando* but now more commonly found as *Fernando* or *Hernando*).

FIVER is the clairvoyant little rabbit in Richard Adams' modern animal classic, *Watership Down* (1972). His name is explained early in the book: "'Fiver?' said the other rabbit. 'Why's he called that?' 'Five in the litter, you know: he was the last—and the smallest'" (Chapter 1). A footnote explains: "There were probably more than five rabbits in the litter when Fiver was born, but his name [in rabbit-speak], *Hrairoo*, means 'Little thousand,' i.e. the little one of a lot or, as they say of pigs, 'the runt.'"

FLICKA is the wild filly in Mary O'Hara's novel *My Friend Flicka* (1941). Her name arose from the remark made by the Swedish ranch hand Gus to Ken, the 10-year-old American boy who chose her from the yearlings on his father's Wyoming ranch:

> "Yee whiz! Luk at de little *flicka*!" said Gus.
> "What does *flicka* mean, Gus?"
> "Swedish for little gurl, Ken —"
> Ken announced at supper, "You said she'd never been named. I've named her. Her name is Flicka."

A movie version of the story was released in 1943.

GARFIELD is the fat cat created in 1977 by the American cartoonist Jim Davis for the comic strip that bears its name. Davis gave the cat his grandfather's middle name.

## 12. Animals in Fiction

GARM is the dog, "two parts bull and one terrier," in Rudyard Kipling's short story "Garm — A Hostage," in *Actions and Reactions* (1909). His name is that of the monstrous hound of Norse mythology who was chained at the gate of hell to watch over the dead. Its own name, spelled GARMR, probably represents Old Norse *harmr*, "grief," "sorrow" (modern English *harm*).

GARRYOWEN is the citizen's "bloody mangy mongrel" in James Joyce's *Ulysses* (1922). Joyce based the dog on a real one of the name, owned by the father of his aunt, Josephine Murray. It was itself presumably named for *Garryowen*, the place near Limerick in Ireland.

GELERT is the wolfhound of medieval legend who is said to have been slain by his master, Prince Llewellyn, for allegedly devouring the Prince's baby son when he had actually killed a wolf that threatened the child. The dog is supposed to be buried at the Welsh village of *Beddgelert*, whose name is popularly interpreted as "Gelert's grave," from Welsh *bedd*, "grave," and his own name. But the village is very likely named for an obscure Welsh saint called *Celert*, a name of uncertain origin, and the story about Gelert is probably an 18th-century invention on the part of an enterprising innkeeper with the aim of attracting visitors to his hostelry.

GERTRUDE is the kangaroo who appears as the colophon (trademark) for the American paperback publishers, Pocket Books. She first appeared in 1939 and was designed by Frank J. Liebermann, who named her for his mother-in-law.

GIDEON is the cat familiar from the Disney animated cartoon *Pinocchio* (1940), based on English translations of the Italian story of 1883 by "Carlo Collodi" (Carlo Lorenzini). His name appears to be based on Italian *il Gatto*, "the Cat," as which he is known in the original.

GODZILLA is the huge lizard who first appeared in 1955 in the Japanese movie *Godzilla, King of the Monsters*, in which he threatened to destroy Tokyo. To English speakers, his name suggests a blend of *god* and *gorilla*. Perhaps significantly, he takes on KING KONG in a sequel, *King Kong vs. Godzilla* (1962). The Japanese title of the original movie was *Gojira*.

GOOFY is the gawky black hound in the Disney animated cartoons starring MICKEY MOUSE (see below). His name expresses his nature, as a stupid or *goofy* animal. He first appeared as DIPPY DAWG (i.e., "silly dog") in the cartoon *Mickey's Revue* (1932). He mostly keeps his English name in non–English languages, although the French know him as *Dingo* ("Screwy"), and the Portuguese as *Papeta* ("Stupid").

GRIP is the pet raven of Barnaby Rudge in the novel of the same name (1841) by Charles Dickens. His name was adopted from that of the author's own raven, itself presumably so called for its firm grasp when tearing at meat.

GRIZABELLA is the scraggy, sexy cat played by Elaine Paige in Andrew Lloyd Webber's musical *Cats* (1981), based on the poems in T.S. Eliot's *Old Possum's Book of Practical Cats* (1939). The name, which does not occur in the original, is doubtless based on French *gris*, "gray," and *belle*, "beautiful."

GRIZZLE is the old gray mare ridden by Dr. Syntax in William Combe's verse satire *The Tour of Dr. Syntax in Search of the Picturesque* (1809). His name describes his color, and he is prominent in the illustrations to the work by Thomas Rowlandson.

GUS is the hero of the poem "Gus: The Theatre Cat" in T.S. Eliot's *Old Possum's Book of Practical Cats* (1939). The poem explains that his name is short for ASPARAGUS, but does not give the source of this name. For a real theatre cat, see BEERBOHM in Chapter 10, page 107.

HARVEY is the six foot rabbit that is invisible to all except the amiable drunkard Elwood P. Dowd, who is the real hero of *Harvey* (1944), by the American playwright Mary Coyle Chase. The play comically debunks psychiatrists, one of whom, Sanderson, asks Elwood why he calls his rabbit thus:

> ELWOOD: (*proudly*) Harvey is his name.
> SANDERSON: Yes; but how do you know that?
> ELWOOD: Ah!—now that's a very interesting coincidence. One night, several years ago, I was walking down Fairfax Street—between Eighteenth and Nineteenth—you know that block?
> SANDERSON: Yes, Mr. Dowd.
> ELWOOD: [...] I started to walk down the street when I heard a voice saying, "Good evening, Mr. Dowd." I turned, and there was this great big white rabbit leaning against a lamp-post. Well, I thought nothing of that! Because, when you've lived in a town as long as I've lived in this one, you get used to the fact that everybody knows your name. So I went over to chat to him. [...] He went on talking, and finally I said: "You have the advantage over me. You know my name—but I don't know yours." And right back at me he said: "What name do you like?" Well, I didn't have to think a minute. "Harvey" has always been my favorite name; so I said: "Harvey." Now this is the most interesting part of the whole thing. He said: "What a coincidence; my name happens to be—Harvey!"

HATHI is the elephant in Rudyard Kipling's *Jungle Books* (1894, 1895). As for many of the other animals, his name is simply the Hindustani word for the animal that he is, in this case *hāthi*, "elephant."

HAZEL is the yearling rabbit in Richard Adams' classic *Watership Down* (1972) who leads an exodus from his warren when his brother FIVER (see above) senses it is about to be destroyed. He is presumably named for the color of his fur or eyes, or from the *hazel* (filbert) bushes near his form. Other rabbits in the novel have similar names, such as BLACKBERRY and BUCKTHORN.

HERVEY is the squinting dog in Rudyard Kipling's story "The Dog Hervey" in *A Diversity of Creatures* (1917). He is at first called HARVEY by a family "committee of names," after *Harvey's Sauce,* but is then renamed HERVEY for Dr. Johnson's famous remark about an acquaintance of his, quoted in James Boswell's *The Life of Samuel Johnson* and noted in 1737: "He was a vicious man, but very kind to me. If you call a dog *Hervey,* I shall love him." Johnson was referring to his friend, a young man of his own age, the Hon. Henry Hervey, third son of John Hervey, 1st Earl of Bristol.

HONEST JOHN is the cunning fox in the Disney cartoon *Pinocchio* (1940), based on the Italian original by "Carlo Collodi" (Carlo Lorenzini), in which he is simply *il Volpe,* "the Fox." His name is intentionally ironic, of course. He is also known as J. WORTHINGTON FOULFELLOW, a mock-pompous name perhaps based on that of the cartoon character *J. Wellington Wimpy.* In French the fox is known as *Grand Coquin* ("Big Rascal"), while in German he is *Fuchs Tunichtgut* ("Fox Ne'er-do-well"). Some languages translate his English name literally, such as Spanish *Honrado Juan* and Portuguese *João Honesto.*

HUEY, DEWEY and LOUIE are the three cartoon ducklings who made their first appearance in Disney's *Donald's Nephews* (1938), this title describing their relationship to DONALD DUCK. Their names are based on arbitrary rhymes. In French they are known as *Riri, Fifi, Loulou,* in German as *Tick, Trick, Track,* in Spanish as *Juanito, Jorgito, Jaimito* ("Johnnie, Georgie, Jamie"), in Italian as *Quo, Qui, Qua* (suggesting quacks), and in Dutch as *Kwik, Kwek, Kwak* (similarly).

HUNCA MUNCA is the mouse in the children's picture book *The Tale of Two Bad Mice* (1904) by Beatrix Potter. The author presumably adopted the name from that of *Huncamunca,* the sweet and gentle (and amorous) daughter of King Arthur and Queen Dollallolla in Henry Fielding's farce *Tom Thumb* (1730). Although derivative, the name is more original than those of the majority of Potter's characters. (For a full listing, see Appendix III.)

JEOFFREY is the cat who is the subject of a section of Christopher Smart's extraordinary poem *Jubilate Agno,* an unfinished work celebrating the Creation and composed between 1758 and 1763. The opening lines are probably the best known: "For I will consider my Cat Jeoffrey./ For he is the servant of the Living God, duly and daily serving him." Two later lines run: "For he knows that God is his Saviour. / For there is nothing sweeter than his peace when at rest." This, against the general religious background of the poem, suggests that Smart had chosen the name (modern *Geoffrey* or *Jeffrey*) for its popular interpretation as "God's peace," properly that of the name *Godfrey*.

JIMINY CRICKET is the cricket familiar from the Disney animated cartoon *Pinocchio* (1940), based on English versions of the Italian original by "Carlo Collodi" (Carlo Lorenzini), in which he is simply *il Grillo Parlante*, "the Talking Cricket." His English name is simply an adoption of the exclamation of surprise, itself a euphemistic form of *Jesus Christ*. His name is similar in other languages, such as French *Jiminy Criquet* and German *Jiminy Grille*.

JIP is Dora Spenlow's little spaniel in Charles Dickens' *David Copperfield* (1850). His name is a short form of GYPSY, a common name for a dog that wanders or is missing (as JIP is on one occasion). For a similar name see ROVER in Chapter 2, page 21.

JOCK is the dog who is the central character in one of the most popular dog stories of all times, Sir Percy Fitzpatrick's *Jock of the Bushveld* (1907), set in the South African goldfields. He is a bull terrier, but his Scottish name is a compliment more to his mother, JESS, than to his nationality. Fitzpatrick himself was Irish. The dog was the smallest and ugliest of a litter of six, and was at first nicknamed "The Rat."

JONATHAN LIVINGSTON SEAGULL is the seagull who is the central character of Richard Bach's cult story (1970) that bears his name. The origin of the name is not given, but "Livingston" seems to hint at the gull's desire to break free from the flock and explore the limits of flying (although of course the famous explorer himself spelled his name with a final -*e*). Other gulls in the story have similar names, such as Fletcher Lynd Seagull, Henry Calvin Gull, Martin William Seagull, and Terrence Lowell Gull.

JUMBLE is the pet dog of the scruffy schoolboy William Brown in the books (1922–70) by Richmal Crompton. He is so named because he is a mongrel, a "jumble" of breeds.

KAA is the python in Rudyard Kipling's *Jungle Books* (1894, 1895). It was devised by the author to represent what he called "the queer openmouthed hiss of a big snake." It is best pronounced unvoiced in this respect.

KASHTANKA is the small red mongrel in Anton Chekhov's short story of the same name (1887). The dog's name derives from Russian *kashtan*, "chestnut," referring to his color.

KIKI-LA-DOUCETTE is the male (*sic*) Angora cat in Colette's *Dialogues de*

*Bêtes* ("Animal Dialogues") (1904), in which he converses with the bulldog TOBY-CHIEN (see below). The cat was based on a real one of this name owned by Colette, with *Kiki* simply a French name of affection and *Doucette* meaning "timid."

KING KONG is the giant gorilla in the famous movie (1933) named for him. His name could simply mean "king" twice over, since the second word also has this sense, if based on Danish *konge,* German *König,* etc. On the other hand the name may simply be a reduplication, like *ding dong* or *Ping Pong.*

LADY is the "brach" or bitch hound mentioned by Hotspur in Shakespeare's *Henry IV, Part I* (1597). Her name is of obvious origin.

LASSIE, famous from the movies, originated as the faithful Yorkshire collie who is the heroine of Eric Knight's short story *Lassie Come Home.* This originally appeared in *The Saturday Evening Post* in 1938 but two years later was expanded into a book. The name is of obvious origin, although in the first movie of the series, screened in 1943 and named for the story on which it was based, the bitch was actually played by a dog named PAL.

LION is Henry Gowan's Newfoundland dog in Charles Dickens' *Little Dorrit* (1857). His name is self-descriptive, but is perhaps intended ironically when he is beaten into submission by his master.

LOVELY is the "Inestimable Dog" mentioned by the gentleman in the Circumlocution Office in Charles Dickens' *Little Dorrit* (1857):

> "What did he call the Dog?"
> "Called him Lovely," said the other gentleman. "Said the Dog was the perfect picture of the old aunt from whom he had expectations. Found him particularly like her when hocussed [drugged]."

LOVEY is the central character in O. Henry's short story "Memoirs of a Yellow Dog" (1917). His obvious name is described as "something of a nomenclatural tin can on the tail of one's self-respect." He was subsequently renamed PETE. Another dog in the story is called TWEETNESS.

LUCKY is the resourceful puppy of PONGO (see below) in Dodie Smith's children's novel *The Hundred and One Dalmatians* (1956). He is so named from the horseshoe pattern of dots on his back.

LUFRA, "the fleetest hound in all the North," is the dog of the outlawed Lord James of Douglas in Sir Walter Scott's poem *The Lady of the Lake* (1810). She is the fond companion of Douglas's daughter Ellen: "They were such playmates, that with name/ Of Lufra, Ellen's image came." The origin of the dog's name is uncertain, but the "loving" connection suggests that it may be in Old English *lufu,* "love."

MAJOR is the prize Middle White boar in George Orwell's satirical novel

*Animal Farm* (1945). He is named for his seniority and authority: "so he was always called, though the name under which he had been exhibited was Willingdon Beauty."

MEHITABEL is the alley cat in the stories and verse by the American writer Don Marquis about *archy and mehitabel* (1927), the cockroach and the cat who lived in the office of the New York *Sun*. (The text always lacks capitals as it was typed by archy who cannot work the typewriter shift key.) The cat bears the name of an Old Testament character mentioned in a genealogy ("And his wife's name was Mehetabel, the daughter of Matred," Genesis 36.39), but perhaps was intended punningly to suggest "me hit a bell." MEHITABEL claims that she once lived in ancient Egypt in the body of Cleopatra, and it could be that she (through Marquis) chose this name for her previous incarnation because of its "feline" resonance: *Cleopatra* not only contains the words *cat* and *leo* ("lion") but is not all that far off *leopard*.

MERRYLEGS is the performing circus dog in Charles Dickens' *Hard Times* (1854). His name is self-descriptive.

MICKEY MOUSE, Walt Disney's creation, who hardly needs any introduction, made his debut in the animated cartoon (the first with sound) *Steamboat Willie* (1928). He was originally supposed to have been named MORTIMER MOUSE, but was called MICKEY MOUSE instead through the influence of Lillian, Disney's wife. Both names are randomly alliterative. In French he is *Mickey la Souris*, in German *Mickey-Maus*, in Spanish *Ratón Mickey* ("Mouse Mickey"), in Italian *Topolino* ("Little Mouse"), in Swedish *Musse Pigg* ("Baby Mouse"), in Japanese *Miki Kuchi* ("Mickey Mouth" [sic]), in Russian *Myshonok Mikki* ("Baby Mouse Mickey"). See also MINNIE MOUSE, below.

MILOU is the white terrier of Tintin in the comic strip about the boy detective created by Hergé in 1929. His name is simply a term of affection. In the English editions of the adventures he is SNOWY.

MINNIE MOUSE is the sweetheart of MICKEY MOUSE in the famous Disney animated cartoons. She was with him from the start, and has a name that, like his, is essentially a random alliteration, although MINNIE also suggests smallness. She mostly keeps her name in other languages, although with minor respellings, such as Italian *Minni*.

MINON is the enchanted cat in Charles Lamb's poem for children, *Prince Dorus* (1811). Her name represents the French generic cat name related to *mignon*, "darling," so more or less corresponds to English PUSSY. See also Chapter 2, page 20.

MISTIGRIS is the cat belonging to Mme. Vauquer in Honoré de Balzac's novel *Le Père Goriot* ("Old Goriot") (1834). His name is a French familiar word for "cat" itself, from *miste,* a form of *mite,* "mite," and *gris,* "gray." (See also Chapter 2, page 20.)

MOBY-DICK is the white sperm whale in Herman Melville's novel of the same name (1851). The whale was perhaps so called after *Mocha Dick,* a white whale that was the subject of a feature by J.N. Reynolds in *The Knickerbocker Magazine* in 1839, itself possibly so called for its gray color, like that of the *mocha* stone, a variety of chalcedony. However, although Melville may have read this account, he does not quote from it in any of the 60-odd "Extracts" about whales and whaling with which he opens the novel, and it is strange that amid all the historical and technical details of whaling that the novel contains, including even an etymological note on the word *whale,* he gives no explanation of the name.

MONTMORENCY is the dog in Jerome K. Jerome's entertaining *Three Men in a Boat* (1889), the tale of three young men who take a rowing holiday on the Thames. The dog has a typically aristocratic name.

MURR is the cat who is the subject of E.T.A. Hoffmann's fictional autobiography *Kater Murr* ("Tom-Cat Murr") (1820). He was based on Hoffmann's own cat of this name, which itself means something like "purrer," from an imitative word found also in German *murmeln,* "to murmur," "to mutter," and so in English *murmur* in turn.

NANA is the Newfoundland dog who, as her name implies, is the children's *nana* (nurse) in J.M. Barrie's *Peter Pan* (1904). The dog was modeled on Barrie's own (male) Newfoundland, LUATH.

NAPOLEON is the large Berkshire boar who is the ruthless leader in George Orwell's satirical novel *Animal Farm* (1946). He obviously takes his name from the famous French emperor. The name is doubly appropriate since in the novel he represents the Russian dictator Stalin.

OLD YELLER is the big, ugly, adopted dog in Fred Gipson's novel of the same name (1956). He is named punningly both for his howl and his color. A sequel (1962) tells of the adventures of his son, SAVAGE SAM.

ORLANDO, "the Marmalade Cat," is the central character in the children's picture books by the Scottish writer Kathleen Hale. He first appeared in *Orlando the Marmalade Cat: A Camping Holiday* (1938), with his wife GRACE and their children, the tortoiseshell PANSY, the white BLANCHE, and the black TINKLE. By the mid–20th century his name had become a household word in Britain. It is properly the Italian personal name corresponding to *Roland,* but was perhaps chosen because it suggests his color through such words as *orange, oriole,* and *or* (gold). It also echoes *marmalade* itself.

OWD BOB is the gray collie and main character in the novel (1898) by Alfred Ollivant of the same name. His name is a Scots form of "Old Bob."

PADDINGTON is the bear who features in the popular children's stories by Michael Bond. He first appeared in *A Bear Called Paddington* (1958), in which he is found by the Brown family on Paddington Station, London. The author was himself living near this station when he first wrote the stories. The name is also suitable for an animal that "pads" along.

PAN is the spaniel who belongs to Bevis in Richard Jefferies' novel, *Bevis: The Story of a Boy* (1882). The story is set almost entirely in the countryside, so that PAN may be intended to evoke the Greek god of fields, woods, shepherds, and flocks. At the same time the dog is a *spaniel*, and his name may derive from this breed. (Compare PONTO below.) It even hints at the com*pan*ion that the dog is to his young master.

PEGASUS, the winged horse, is one of the best known animals of classical mythology. He was born of the blood of Medusa when she was killed by Perseus. Medusa was pregnant by Poseidon at the time, so the horse had the god of the sea as his father. The popular derivation of his name is from Greek *pēgē*, "spring," said to refer to the springs (*pēgai*) of Oceanus, near which Medusa was killed. As so often, however, the name is probably pre-Greek in origin. However, at least two springs in Greece were attributed to a stamp on the ground of the horse's foot, the most famous being the *Hippocrene*, "horse spring," on Mount Helicon.

PERDITA is the rescued stray in Dodie Smith's popular children's book *The Hundred and One Dalmatians* (1956). Her name was adopted from that of the daughter of Hermione and Leontes in Shakespeare's *The Winter's Tale*, as the narrative specifically indicates:

> "We'll call her Perdita," said Mrs. Dearly, and explained to the Nannies that this was after a character in Shakespeare. "*She* was lost. And the Latin word for lost is 'perditus.'"

PLUTO is the lanky, loopy hound in the animated cartoons by Walt Disney. He first appeared as an anonymous bloodhound in *The Chain Gang* (1930). His name is probably arbitrary, although properly suggests the god of the Underworld in Greek mythology.

PONGO is the father of 15 puppies in Dodie Smith's children's novel *The Hundred and One Dalmatians* (1956). Describing the dog, the author comments thus on his name: "He remembered that Mr. Dearly had named him 'Pongo' because it was a name given to many Dalmatians of those earlier days when they ran behind carriages." The origin of the name itself is obscure, although it was sometimes used in the 19th century for the dog (more commonly known as TOBY) in the Punch and Judy puppet show.

PONTO is a dog found in various well-known literary works. One of his earliest occurrences is in Tobias Smollett's epistolary novel, *The Expedition of Humphry Clinker* (1771), where he is mentioned humorously in a letter from Jeremy Melford to his friend Sir Watkin Phillips:

> I cannot, however, approve of his drowning my poor dog Ponto, on purpose to convert Ovid's pleonasm unto a punning epitaph, — *deerant quoque littora Ponto:* for, that he threw him into the Isis, when it was so high and impetuous, with no other view than to kill the fleas, is an excuse that will not hold water.

(The Latin quotation literally translates as "The shores also failed the Black Sea [*Pontus*]," but could be punningly understood to mean "The banks also failed Ponto.") More familiar is the PONTO who is Mr. Jingle's sagacious sporting dog in Charles Dickens' *The Pickwick Papers* (1837). His name undoubtedly relates to that of his breed, a *pointer*. A third PONTO appears in R.S. Surtees' *Mr. Sponge's Sporting Tour* (1852), where he is described as "a great fat, black-and-white brute, with a head like a hat-box, a tail like a clothes'-peg, and a back as broad as a well-fed sheep's" who serves as "pointer, house-dog, and horse to Gustavus James." In more recent literature, the name has been popularized by the comic history of the little boy Jim who was eaten by a lion when he let go of his nurse's hand at the zoo, in Hilaire Belloc's *Cautionary Tales for Children* (1907): "'Ponto!'" he cried, with angry frown,/ 'Let go, Sir! Down, Sir! Put it down!'"

PRACTICAL PIG, FIFER PIG, and FIDDLER PIG are the familiar trio of pigs who made their debut in the Disney "Silly Symphony" animated cartoon *Three Little Pigs* (1933). Their names describe their attributes, with PRACTICAL PIG being the serious, thoughtful mason, while light-hearted FIFER PIG and FIDDLER PIG play the flute and violin respectively. It seems strange that no snappier English name was devised for them. In French they have the nonsense names *Naf-Naf, Nif-Nif, Nouf-Nouf,* and in Dutch, similarly, are known as *Knor, Knir, Knar*. Other languages translate the English names, such as German *Schweinchen Schlau, Pfeifer, Fiedler* ("Pigling Smart, Fifer, Fiddler") or Spanish *Práctico, Flautista, Violinista*.

PYEWACKET is the Siamese cat who is the "familiar" of Gillian Holroyd in John van Druten's play, *Bell, Book and Candle* (1950). The name was adopted from that of one of the "familiars" of a witch mentioned in Matthew Hopkins' *Discovery of Witches* (1647) (see Chapter 11, page 135, for further details and more names of this type). In the actual record of the trial of this witch, the cat's name was given as PYNEWACKET. The meaning of this is obscure, as it is for many other names of "familiars."

Van Druten's play was later made into a movie (1958), with Kim Novak playing Gillian Holroyd.

RAKSHA is the mother wolf who suckled the baby Mowgli in Rudyard Kipling's *Jungle Books* (1894, 1895). Her name represents Hindustani *raksha*, "care," "protection."

REGULA BADDUN is a racehorse in Rudyard Kipling's short story "The Broken Link Handicap" in *Plain Tales from the Hills* (1888). The horse has an obviously punning name which, however, was intended as a reference to the heartless and unscrupulous Anglo-Indian lady Mrs. Reiver, who appears in this and other stories in the book.

REX is the fox terrier, a "soft fat pup," in the short story of the same name in D.H. Lawrence's *The Mortal Coil and Other Stories* (1924). The dog had a black spot at the base of his spine, so that at first the name SPOT was considered for him:

> But that was too ordinary. It was a great question, what to call him. "Call him Rex the King," said my mother. [...] We took the name in all seriousness. [...]
>
> We thought it was just right. Not for years did I realize that it was a sarcasm on my mother's part. [...]
>
> It wasn't a successful name, really. Because my father and all the people in the street failed completely to pronounce the monosyllable Rex. They all said Rax. And it always distressed me. It always suggested to me seaweed, and rack-and-ruin. Poor Rex!

RIKKI-TIKKI-TAVI is the mongoose in Rudyard Kipling's *Jungle Books* (1894, 1895). According to the story in which he appears (and which is named for him) his name represented "his war-cry, as he scuttled through the long grass," this being "Rikk-tikk-tikki-tikki-tchk!" Kipling may also have had in mind Hindustani *ṭikṭikki*, "lizard."

ROOSEVELT is a puppy in Hugh Walpole's novel *Head in Green Bronze* (1938), set in Los Angeles. He was named for Franklin D. *Roosevelt*, American president from 1933 to 1945, but blamed by the public in 1937 for labor difficulties and criticized by businessmen for his economic policies.

> Isabella insisted that it should be called Roosevelt.
> "Why?" asked William.
> "Well, I think he's the most wonderful man in the world [and] now, when people are turning against him [...] one has to stand up for him and come right out into the open."
> "I don't see," said William, "that calling the puppy Roosevelt is coming out into the open."
> "It's a kind of demonstration. After all, isn't the puppy the sweetest thing in the world?"

"I don't think," said William sulkily, "that Roosevelt would like anyone to call him the sweetest thing in the world. He isn't at all that kind of man."

ROSINANTE is Don Quixote's sorry steed in the satirical romance by Cervantes. His name means literally "nag before," from Spanish *rocín,* "hack," "nag," and *antes,* "before." After deliberating four days what to call him, Don Quixote reached his decision:

> Al fin le vino á llamar Rocinante, nombre, á su parecer, alto, sonoro y significativo de lo que había sido cuando fué rocín, antes de lo que ahora era, que era antes y primero de todos los rocines del mundo. [He ended by calling him Rocinante, a name, it seemed to him, that was lofty, sonorous, and indicative of what he had been when he was a hack, before what he was now, and of how he was the first and foremost of all hacks in the world. Miguel de Cervantes, *Don Quijote,* Part One, Book I, Chapter 1, 1608.]

ROSWAL is the staghound of Sir Kenneth, the Scottish crusader in Walter Scott's *The Talisman* (1825). His name may be a combination of the Germanic words *hros,* "horse," and *wald,* "wood," referring to his role accompanying hunters in the forest. The dog himself was based on Scott's beloved MAIDA (see Chapter 10, page 122).

ROWF is the big black mongrel who scarcely survives experiments in Richard Adams' novel *The Plague Dogs* (1978). His name presumably represents a bark.

RUKSH is the "bright bay, with lofty crest" owned by Rustum in Matthew Arnold's poem "Sohrab and Rustum" (1853). He and his name were based on RAKHSH, the horse who belonged to the legendary Persian hero Rustum in the 11th-century epic poem *Shah-nameh* ("Book of Kings"). His name derives from a Persian root word meaning "light," "shining."

SALAR is the salmon in Henry Williamson's tale of the countryside, *Salar the Salmon* (1935). The author explains the name as meaning "leaper." If this is the case, both *Salar* and *salmon* can be linked with Latin *salire,* "to leap" and so with English *salient* and *somersault.*

SAN FRANCESCO is the tramp's dog adopted by Peter and Thomas in Rose Macaulay's novel *The Lee Shore* (1912). Peter names the dog: "Shall we call you San Francesco, because you like disreputable people and love your brother, the sun, and keep company with your little sisters, the fleas? Very good, then."

SCOOBY-DOO is the great Dane created as a television cartoon character in 1969 by Hanna-Barbera productions. His name seems arbitrary, but has a suggestion of *scoot,* as appropriate for the cowardly canine that he is.

SCRAPS is the little white dog who costarred with Charlie Chaplin in the movie *A Dog's Life* (1918), for which he was billed as a "thoroughbred mongrel." His name evokes his small size, the leftovers he scrounges for food, and the fights he gets involved in. The dog who played him was actually BROWNIE, so named for the big brown patch over his left eye.

SHERE KHAN is the tiger who demands the "man-cub" Mowgli in Rudyard Kipling's *Jungle Books* (1894, 1895). His name means "chief tiger," from Hindustani *sher*, "tiger," and *khān*, "chief." *Sher Khan* was actually the title given to the Mogul emperor Sher Shah (1486–1545) after he had killed a tiger. In Kipling's story, SHERE KHAN is also killed, although not by a human.

SILVER is the Lone Ranger's white stallion (hence his name) in the American West adventures that began as a radio series in 1933 and that progressed to book, comic strip, movie, and television versions. The Lone Ranger, a Robin Hood–style masked adventurer, used special silver bullets, as if to mirror the name of his mount, and his cry "Hi-yo, Silver, Awa-a-ay!" is still a familiar catchphrase.

SILVER BLAZE is the racehorse who disappears in Sir Arthur Conan Doyle's Sherlock Holmes story *The Adventures of Silver Blaze* (1892). The horse's name refers to the distinctive white blaze on his face.

SIRIUS is the dog born with the mind of a man who is the central character in the science fiction novel (1944) by Olaf Stapledon named for him. The name is naturally fitting for a "dog star."

SKULKER is the bulldog in Emily Brontë's novel *Wuthering Heights* (1847) who seizes 12-year-old Cathy Earnshaw's ankle when she and Heathcliff are fleeing from Thrushcross Grange after peering through the window at the Linton family. His name means "one who skulks," i.e., lies in hiding. The name is starkly descriptive, and is matched by those of other dogs in the novel, such as GNASHER and THROTTLER. The latter is SKULKER's son.

SNARF is the dog who features in the drawings of the American cartoonist George Booth (born 1926) in *The New Yorker*. His creator explains how he came to give the name:

> After I created my Bull Terrier several years ago, a gentleman came to my front door with his dog, Snarfi Sue. I liked the name Snarf, and the man gave me permission to use it. It strikes me as a combination of *snarl* and *arf*, and I chose it because *it feels right*. Originally I drew an ornery mutt which was no special breed. People wrote to *The New Yorker* asking whether it was an English Bull Terrier. I began improving his breeding until he is now a full-blooded English Bull Terrier [quoted in Carrie Shook and Robert L. Shook, *What to Name Your Dog;* see Bibliography].

SNARLEYYOW is the dog hero of Captain Frederick Marryat's novel (1837) of the same name, subtitled *The Dog Fiend*. He is introduced thus: "The name of this uncouth animal was very appropriate to his appearance, and to his temper. It was Snarleyyow." An artillery horse named SNARLEYOW appears in the poem of this name in Rudyard Kipling's *Barrack-Room Ballads* (1892).

SNITTER is the black and white fox terrier who is the companion of ROWF (see above) in Richard Adams' novel *The Plague Dogs* (1978). His name suggests *snitcher*, "one that snitches," i.e., steals.

SNOOPY is the famous beagle created by Charles M. Schulz for his comic strip *Peanuts*. He made his first appearance in 1950 and has a name appropriate for a breed of dog that *snoops*, i.e., noses things out. He was based on a real dog called SPIKE that Schulz had owned as a boy.

SNOWBALL is the idealist pig driven out by NAPOLEON (see above) in George Orwell's satirical novel *Animal Farm* (1946). His name ostensibly refers to his pale color, but is also appropriate for the character who in the novel represents Trotsky, the Russian revolutionary ousted by Stalin.

SREDNI VASHTAR is the caged ferret owned by 10-year-old Conradin in the story of the same name in *The Chronicles of Clovis* (1911) by "Saki" (H.H. Munro). The boy "spun the beast a wonderful name" and regarded the animal as "a god and a religion." Munro doubtless invented the name arbitrarily, although *Vashtar* suggests a form of the biblical female name *Vashti*, from the Persian meaning "beautiful."

STENTERELLO is Christina Light's poodle, which follows her everywhere, in Henry James' novel *Roderick Hudson* (1876). He is named for the masked personage found in Florentine comedy from the late 18th century. The dog is intelligent yet cowardly, like this character.

STUMAH is the dog owned by the Highland chief Roderick Dhu in Sir Walter Scott's poem *The Lady of the Lake* (1810). His name is of uncertain origin, perhaps Celtic rather than Old English. It may be related to Gaelic *stumpach*, "stumpy," implying a stocky animal.

SWEETHEART is one of the three dogs mentioned by the distraught king in Shakespeare's *King Lear* (1605) (see TRAY). Her name is of obvious meaning.

SYLVIO is Maria's faithful dog in Laurence Sterne's novel *A Sentimental Journey* (1768), which she had got in place of her faithless (unnamed) goat. The story leads us to believe that the dog was named for Maria's equally faithless lover. The reader is not specifically told his name, but Maria has kept a handkerchief of his with "an *S* mark'd in one of the corners."

TACHEBRUNE is the horse of Ogier the Dane in the French medieval romances about Charlemagne. The name means "brown patch," from French *tache,* "patch," and *brune,* "brown."

TARKA is the otter in Henry Williamson's popular story *Tarka the Otter* (1927). According to the author, the name is one traditionally used for otters by local people on Exmoor, where the story is set, and means either "Little Water Wandering" or "Wandering as Water." If this is so, the name appears to be related to *otter* itself, with the root element of both words seen in English *water.*

TARTAR is Shirley Keeldar's dog, "of a breed between mastiff and bulldog," in Charlotte Brontë's novel *Shirley* (1849). He has an appropriately "tough" name. He was based on KEEPER, the dog of Charlotte's sister, Emily (who was herself the model for Shirley).

TEEM is the little French dog who hunts for truffles in Rudyard Kipling's short story (1935) of the same name. Although not explicitly stated as such, his name is clearly meant to represent a French pronunication of "Tim."

TEST is the dog who is the subject of one of the essays entitled "Popular Fallacies" in Charles Lamb's *Essays of Elia* (1820–23): "XIII. That You Must Love Me and Love My Dog." The name is explained in the following dialogue:

> "What is this confounded cur? He has fastened his tooth, which is none of the bluntest, just in the fleshy part of my leg."
> "It is my dog, sir. You must love him for my sake. Here, Test-Test-Test!"
> "But he has bitten me."
> "Ay, that he is apt to do, till you are better acquainted with him. I have had him in three years. He never bites me." [...]
> "But do you always take him out with you, when you go a friendship-hunting?"
> "Invariably. 'Tis the sweetest, prettiest, best conditioned animal. I call him my *test*—the touchstone by which to try a friend. No one can properly be said to love me, who does not love him."

TOBY is the dog in the traditional Punch and Judy puppet show. The show itself evolved from the Italian *commedia dell'arte,* or improvised comedy of the 16th century, in which the equivalent of Punch was *Pulcinella.* The origin of TOBY, however, is uncertain. He is very probably an English addition to the show, and is first recorded some time between 1810 and 1850. He seems to have made his debut as a live dog, and bore the name either because he had it already, or by association with the biblical story of *Tobias* and the Angel, which was a favorite subject for puppet plays. In this, Tobias and the Angel were accompanied by a dog, who had no name in the original story (see Chapter 11, page 135).

TOBY-CHIEN is the highly strung bulldog in Colette's *Dialogues de Bêtes* ("Animal Dialogues") (1904), in which he converses with KIKI-LA-DOU-CETTE (see above). His name simply means "Toby dog." (See TOBY above.)

TOM and JERRY are the famous cat-and-mouse team in the Hanna-Barbera animated cartoons. They made their debut in *Puss Gets the Boot* (1940). Hanna and Barbera named the characters on the basis of hundreds of suggestions submitted by studio employees in a contest staged at MGM. In the screen debut mentioned, TOM was actually called JASPER. The names are those of an earlier pair: the two chief characters in Pierce Egan's comic *Life in London* (1820), subtitled *The Day and Night Scenes of Jerry Hawthorn Esq. and His Elegant Friend Corinthian Tom*, two rakish "men about town" of their day. TOM is of course an appropriate name for a cat in any case.

TOP CAT, or T.C. for short, is the famous animated cartoon cat created by the Hanna-Barbera team in 1961. His name is obviously patterned on *top dog* as a term for a person in a superior position (literally a *dog* who is on *top* in a fight). He is known as BOSS CAT in Britain, since "Top Cat" there already existed as the registered name of a brand of pet food.

TOTO is Dorothy's little dog in Frank Baum's famous children's novel *The Wonderful Wizard of Oz* (1900). His name is not uncommon and reflects the French children's word for a dog, *toutou*.

TRAY was long a near-generic name for a sporting dog. It famously occurs for one of the three dogs mentioned by the distraught king, in his madness, in Shakespeare's *King Lear* (1605): "The little dogs and all, Tray, Blanch, and Sweet-heart, see, they bark at me." The name has been popularly derived from either German *tragen*, "to carry," "to bear," as if the dog were a kind of messenger, or Spanish *traer*, "to get," "to fetch." It actually means "true," "faithful," from a word related to archaic *trow* and *troth* and modern *true*, so is virtually synonymous with FIDO (see Chapter 2, page 21). The name has been promoted by other dogs in literature from Shakespeare's day on. The poet Matthew Prior (1664–1721) has it in his witty verse dialogue *Alma, or the Progress of the Mind* (1718):

> The sport and race no more he minds;
> Neglected Tray and Pointer lie:
> And covies unmolested fly.

(POINTER is here the name of an individual dog, not the familiar breed.) Prior's near contemporary, the poet John Gay (1685–1732), has a verse devoted to a dog of the name in which the literal meaning is evident:

> My dog (the trustiest of his kind),
> With gratitude inflames my mind;
> I mark his true, his faithful way,
> And in my service copy Tray.

A dog of the name also appears in Thomas Campbell's poem *The Harper* (1799):

> On the green banks of Shannon, when Sheelah was nigh,
> No blithe Irish lad was so happy as I;
> No harp like my own could so cheerily play,
> And wherever I went was my poor dog Tray.

The name was particularly popularized, especially among children, by the English translation (1848) of Heinrich Hoffmann's collection of "cautionary tales," *Strewwelpeter* (1845), which has the lines:

> The trough was full, and faithful Tray
> Came out to drink one sultry day.

(In the German original, the dog's name is *Treu,* "true.") The American songwriter Stephen Foster further promoted the name in his song "My Old Dog Tray" (1853), and it subsequently appears as that of Alice's dog in George du Maurier's novel *Trilby* (1894). In the latter, he is introduced as a dog "whose name, like that of so many dogs in fiction and so few in fact, was simply Tray." More recently, in *If Dogs Could Write* (1929), the English essayist E.V. Lucas comments: "That name, which appears once to have been almost the only one that a dog ever had, is obsolete. I doubt if any dogs are called Tray now. I have never met one." (He mistakenly derives the name from Old English *trog*—in his rendering, *tryg*—"trough," perhaps under the subconscious influence of the familiar lines from *Strewwelpeter* above.)

TREBIZOND is the gray mount of the knight Guarinos in the French medieval romances about Charlemagne. The name was adopted from that of the ancient Greek colony that is now the Turkish port of *Trabzon*. Its own name has been associated with Greek *trapeza,* "table."

WANDA is the angel fish in the John Cleese movie *A Fish Called Wanda* (1988), named for the principal female character, American jewel thief posing as law student *Wanda* Gershwitz, played by Jamie Lee Curtis. Her own name apposes that of *Wendy* Leach, wife of the principal male character, barrister Archie Leach.

WHISKER is the pony of Mr. and Mrs. Garland in Charles Dickens' *The Old Curiosity Shop* (1840). His name refers to his *whisking* tail.

WHITE FANG, with his self-descriptive name, is the part wolf, part dog who is the central character in Jack London's novel of the same name (1906). The novel, one of the best-known dog stories in the world, was written as a sequel to *The Call of the Wild* (1903), in which the main character is the dog BUCK.

WHITE SURREY, in Shakespeare's *Richard III* (1592), is the horse that the king prepares for the fatal battle of Bosworth Field: "Saddle white Surrey for the field tomorrow." He was presumably named for the Earl of *Surrey*, son of the Duke of Norfolk, who appears in the play. He supported the king just as the horse literally supported him.

WILBUR is the small friendly pig who is befriended by CHARLOTTE the spider (and saved by her from the usual fate of nice fat pigs) in E.B. White's children's novel *Charlotte's Web* (1952). The name is given the pig by the little girl Fern, his owner, as "the most beautiful name she could think of."

WINNIE-THE-POOH is the bear in A.A. Milne's famous children's story (1926) of the same name. Both parts of the name are based on those of real animals. WINNIE was an American black bear cub, a regimental mascot, who came to the London Zoo in 1914 and who remained there until 1934. POOH was the name of a swan encountered on holiday by Milne's young son, Christopher Robin, who also appears in the book. In French *Winnie-the-Pooh* is usually known as *Winnie l'Ourson* ("Winnie the Little Bear"), in German as *Winnie Puuh,* in Portuguese as *o Ursinho Puff* ("the Little Bear Puff"), and in Russian as *Vinni-Pukh,* the second word of his name here fortuitously coinciding with Russian *pukh,* "fluff."

YOGI BEAR is the popular bear in the Hanna-Barbera television cartoons. He made his first appearance in 1961, and is said to have got his name from the famous American baseball player *Yogi Berra* (real name Lawrence Berra) (born 1925), whose reputation was by then well established.

# APPENDIX I : BREED NAMES

The names of well-known breeds of animals, especially dogs, cats, and horses, are of interest, in particular since in certain cases the name evolved from that of a single animal. In some instances the word "breed" applies only loosely, so that the group name properly refers to a variety or type, rather than to a distinct, established and registered breed. The name is not capitalized unless it involves a proper name (personal name or placename).

**Abyssinian**  The shorthair breed of cat is named for its African origin. The forerunner of the breed is generally reckoned to be the cat brought to England from Abyssinia in 1868 by a Mrs. Barrett-Lennard. For some time the breed was known as a "hare cat" or "rabbit cat," from the resemblance of the cat's fur to that of a hare or rabbit. It was even erroneously believed that the breed had originated from the cross-mating of a cat and a rabbit, a biologically impossible feat.

**affenpinscher**  The dog, related to the Brussels griffon, is also known as the "monkey dog," from its alert-looking, monkey-like face, and this characteristic is reflected in the breed name, which derives from German *Affe,* "monkey" (English "ape") and *Pinscher,* "terrier" (compare **Doberman pinscher**).

**Afghan hound**  The dog was originally a native of Afghanistan, as a member of the greyhound family.

**Airedale**  The terrier, the largest of its breed, takes its name from the valley of the Aire River, Yorkshire, England, where it was originally bred in the mid–19th century as a cross between an otterhound and a black and tan wire-haired terrier.

**Akita**  The dog is an old Japanese breed, first imported to Britain in the 1930s. It takes its name from the district of Akita, some 300 miles north of Tokyo, where it originated.

**Alsatian** In the early years of World War I the dog was familiar as an "Alsatian wolfhound." Yet this is a misnomer, since the breed does not come from Alsace, nor does it contain a wolf strain. The name was adopted in order to avoid the then undesirable association of "German," as in its present commonly recognized name of **German shepherd**. However, "Alsatian" is still popularly in use for the breed, and is officially registered as a name at the Kennel Club. The dog is also sometimes known simply as a "police dog," from its work with the police force.

**American quarter** *see* **quarter horse**

**American saddle** *see* **saddle horse**

**Andalusian** The breed of horse is a cross between the Iberian (Spanish) horses and the Barb, and takes its name from that of the province in southern Spain where it arose.

**Anglo-Arab** The breed of horse is a composite one, as its name indicates, and derives from a blend of two of the purest breeds, the Arabian and the thoroughbred.

**Angora** The longhair breed of cat originally came from Turkey and is named for that country's capital, now normally known as Ankara. The name "Angora" has been replaced by "Persian" in Britain, although in the United States the original Angoras are now known as "Turkish Angoras." The same city gave the name of two other animals with soft, long hair: the Angora goat and the Angora rabbit. These two animals, and especially the wool of the rabbit, were also known as "Angola," a geographically misleading name in modern terms.

**Appaloosa** The breed of horse, with its distinctive spotted rump, is of American origin and is said to take its name from the horses bred by the Nez Percé Indians in the Palouse country of central Idaho and eastern Washington. The name itself is a local tribal one translated into French, and means "grassy place" (compare modern French *pelouse,* "lawn").

**Appenzell** The "Appenzell mountain dog," to give it its full name, is a Swiss breed name for the canton of Appenzell in northeastern Switzerland, where it was originally bred before being introduced to Britain in the 1930s.

**Arabian** The well-known breed of riding horse is one of the oldest and purest in the world, and obviously has a name that denotes its North African origin. Every registered English thoroughbred has Arabian blood in its veins. [Note: "Arab" is the usual British designation.]

**Balinese** The longhair breed of cat is somewhat similar in appearance to a Siamese. Its name originates from the island of Bali, Indonesia, although it is uncertain when the cats themselves were first exported from that country.

**Barb** The breed of horse is similar to the Arabian but not so spirited. It originated in Morocco and Algeria, northwestern Africa, from the extensive but now historic region known as Barbary. The English form of the name is not an abbreviation but an adoption of French *barbe*, "of Barbary."

**basenji** The breed of dog, famous for being "barkless," comes from North Africa, and its name is a Bantu word meaning simply "native." Grammatically, the word *basenji* is the plural of *mosenji*.

**basset hound** The name of the breed is French, ultimately relating to French *bas*, "low," referring to the dog's low body, on short legs.

**beagle** The breed of dog dates back to at least the 15th century, and the name itself is probably French in origin, from *beegueule*, "with gaping mouth," from Old French *beer*, "to gape," and *gueule*, "throat" (related to English *gullet*). If this origin is correct, it refers to the dog in full cry as a hunting dog. However, the modern French word for the dog is the same as the English, and is adopted from it.

**Bedlington terrier** The precise origin of the breed is uncertain, although it certainly developed in Northumberland, northeastern England, where it was recorded in the early 19th century. The name comes from the town of Bedlington, north of Newcastle upon Tyne.

**Birman** The longhair breed of cat is similar to the Siamese. It is relatively new in Britain, having been imported only in the 1960s. The name actually means "Burmese," as it was in Burma that the cat originated. The spelling of the name derives from the name of the country still current in a number of non–English languages, such as French *Birmanie*, German *Birma*, Italian *Birmania*, etc. (The country's official name is now Myanmar.) See also **Burmese**.

**bloodhound** The breed was introduced to England by the Normans in the 11th century as a hunting dog, that is, one trained to "seek blood." The modern French name for the breed is *limier*, literally "dog on a lead," from a word related to modern English *ligament* and *liana*.

**Border terrier** The dog originally came from the so called Border counties, that is, the Cheviot Hills on the English-Scottish border, where it was known in the 17th century.

**borzoi** The sleek racing hound, first seen in England in the 1870s, derives its name from Russia, its country of origin, where it was used as a fast-running dog for hunting wolves. Russian *borzoy* means "swift."

**Boston terrier** The American breed is named for Boston, Massachusetts, where it was exported from England in the 1870s, as a cross between an English and a French bulldog.

**Bouvier des Flandres** The French breed has a name meaning literally "cowherd of Flanders," since the dog was originally a cattle drover used by Flemish farmers.

**boxer** The name refers to the dog's fighting habits as a "boxer," or more precisely to its manner of "boxing" with its forepaws when about to fight. The first of the breed was a dog called FLOCKI, a cross between the bitch of a bull-fighting breed and a bulldog named TOM shown at Munich, Germany, in 1895.

**Brabançon** The breed of draft horse has a French name meaning "of Brabant," referring to the region of Belgium where it originated.

**Briard** The French sheepdog, otherwise known as a "berger de Brie," takes it name from Brie, its region of origin (where the cheese also comes from).

**Brittany spaniel** The dog, also known by its French name of "Epagneul Breton," takes its name from Brittany, northwest France, its district of origin. It is colloquially known as a "Brit."

**bull mastiff** As the name implies, the dog is a crossbreed of a bulldog and a mastiff. It is still also sometimes known as a "keeper's night dog," describing its role as a guard dog. Although familiar enough, the name has oddly not yet found its way into even the latest edition of the *Oxford English Dictionary*, although it appears in almost every other modern dictionary.

**bulldog** The dog derives its name from the former sport of bull-baiting, in which it was regularly employed from as early as the 13th century down to 1835, when (at least in England) the "sport" was abolished. The dog happens to have a bull-like head, which reinforces the breed name.

**Burmese** The shorthair breed of cat originated in Burma, as its name indicates. It was exported to the United States in 1930, when a female brown cat of the as yet unnamed breed was brought from Burma and was mated to a Siamese. British Burmese cats owe their origin to a litter born in the early 1950s to a male and female (queen) cat brought over from the States. The name should not be confused with that of the **Birman**.

**cairn terrier** The small working breed of terrier was developed in the western Highlands of Scotland. Its name is said to derive from its former use in hunting foxes, badgers, wildcats and the like among cairns (heaps of stones used as landmarks along mountain paths, especially in lying snow).

**Cavalier King Charles spaniel** The Cavalier, as it is known for short, is a breed dating back no earlier than the 1920s, when the toy spaniel was developed on the lines of the King Charles spaniel, itself often portrayed in paintings of the time of Charles II. Such pictures were characteristically of Cavaliers (Royalist supporters of Charles I, father of Charles II, in the English Civil War), hence the name. The dog of the breed is now distinctively larger than the King Charles itself.

**Chesapeake Bay retriever** The breed is related to the modern Labrador, and originated from two English puppies, CANTON and SAILOR, who were shipwrecked on the Chesapeake Bay coast, Maryland, in 1807. The familiar name of the breed is "Chessie."

**Chihuahua** The breed of dog is of Mexican origin, developed in the United States, and is said to go back ultimately to the sacred dogs of the Aztecs. Its name is that of the state (and city) in Mexico from which it originally came. The colloquial short name of the breed is "Chi."

**chinchilla** The breed of longhair cat is believed to have evolved by crossing the silver tabby with the blue tabby. It is famous for its silver-colored coat, and takes its name from the squirrel-sized rodent, *Chinchilla laniger*, that itself has soft, pearly-gray fur and that is native to the mountains of Peru and Chile. Its own name is a Spanish corruption of its native name. The chinchilla rabbit takes its name from the same rodent.

**chowchow** The chow (to give it its more familiar, shorter name) is an ancient Chinese breed of dog, with a name deriving from a Chinese dialect word related to Cantonese *kau*, "dog."

**Cleveland bay** The horse of this name claims to be the oldest "established" breed in England. Its precise origin is uncertain, but the name itself refers to that region of northeastern England (now a county) where it is indigenous. The second word of the name relates to its color, which is either bay (reddish-brown) or bay brown.

**Clumber spaniel** The breed of spaniel was developed in about 1770 at Clumber Park, seat of the Duke of Nottingham near Sherwood Forest, Nottinghamshire, England, and is named for its place of origin.

**Clydesdale** The heavy and powerful breed of carthorse derives its name from the region of southern Scotland where it originated in the mid–18th century. Its precise place of origin is the wide valley of the upper Clyde River, corresponding to the historic county of Lanarkshire.

**cocker spaniel** The spaniel evolved from the field spaniel and springer spaniel in the 19th century, when it was trained to flush up woodcock in the sport of "cocking," that is, shooting woodcock. Hence its name.

**collie** The well-known Scottish breed of dog, also known as the "Scotch Colley," probably takes its name from the Scots word *coll*, "coal," referring to its former black coat. An individual dog of the name is mentioned in Chaucer's *Canterbury Tales* ("The Nun's Priest's Tale"): "Ran Colle our dogge, and Talbot, and Gerlond." The link between the name of this 14th-century literary dog and that of the Scottish breed is uncertain.

**Connemara** The breed of pony originated in the barren coastal region of western Ireland for which it is named. Until relatively recently it was found there in a near-wild state, with an existence something like that of the present Dartmoor or Exmoor pony.

**coonhound** The American breed of dog is so named from its use for hunting raccoons. The name has been current since the 19th century.

**corgi** The former cattle dog, in full known as "Welsh corgi," was long the sole breed in Wales. Its name is Welsh and means literally "dwarf dog," alluding to its small size, from *corr,* "dwarf," and *gi,* a mutated form of *ci,* "dog." The breed became common in Britain and the United States only in the 20th century. In Britain it is familiar from being the breed favored by Queen Elizabeth (in the Pembrokeshire variety).

**criollo** The pony of this name is a native of South America, deriving from the original Arab and Barb strains brought there through Spain at the time of the conquest (16th century), when the Spanish cavalry were perhaps the finest in Europe. The name is Spanish for "Creole," that is, relates to a native breed. The two famous horses GATO and MANCHA (see page 117) were criollos.

**dachshund** It was only in the 19th century that the breed of dog became known outside its native Germany, where it was used for hunting or digging out badgers. Hence its name, meaning "badger dog," from German *Dachs,* "badger," and *Hund,* "dog." Its German provenance made the breed highly unpopular in Britain in World War I, and although it was subsequently accepted, it received the opprobrious nickname "sausage dog," partly with reference to its long, brown body, partly through its associations with the German sausage (wurst). It is sometimes known in English by its alternate German name of *Teckel,* which is the standard name for the breed in French.

**Dalmatian** The dog was originally known in England as "coach dog" or "carriage dog," from its use as a guard dog running with mail coaches. Later it was also known as a "Danish dog" or a "Lesser Dane" (see **great Dane**). It is named for Dalmatia, the region successively in Austria, Italy, Yugoslavia, and currently Croatia. Its actual place of origin remains uncertain, however. The breed's colloquial name is "Dally."

**Dandie Dinmont** The dog is a breed of terrier from the Scottish borders. Its name is of literary origin, and derives from the pairs of dogs called MUSTARD and PEPPER owned (and bred) by Dandie (i.e., Andrew) Dinmont, the "store farmer" in Walter Scott's novel *Guy Mannering* (1815).

**Dartmoor** The small, strong, hardy breed of pony is still found in a practically wild state on the bleak moor in southern Devon, southwestern England, for which it is named. Not far to the north the **Exmoor** pony is similarly found.

**deerhound** As its name indicates, the breed of dog, also known as "Scottish deerhound," was used for hunting deer.

**Doberman pinscher** The Doberman, as it is commonly called, is a German dog that originated in 1890 as a terrier strain bred by Ludwig Dobermann, of Apolda, Thuringia, from the bitch SCHNUPP ("Snoopy"). "Pinscher" is a German word for a breed of hunting dog, perhaps itself so named for its place of origin, Pinzgau, Austria.

**elkhound** The breed of dog originated in Scandinavia where it was used for hunting elks before it was introduced to Britain in the 1870s.

**English setter** The distinctive breed of setter was originally known in England in the 16th century. It is now found internationally.

**Exmoor** The pony is the descendant of the native British wild horse, and is named for the high, bleak moorland in western Somerset and northern Devon, in the southwest of England, where it is still found running wild. Not far to the south is the **Dartmoor** pony.

**fox terrier** As its name implies, the dog was originally used as a terrier for unearthing foxes. The wire-haired type is still occasionally used for this purpose today, although more often it is a pet or household dog.

**foxhound** One of the most familiar hunting breeds, with a name of obvious origin. The dog is still mainly used for hunting, rather than being a housedog or pet, and is kept in packs in the kennels of a particular hunt.

**German shepherd** The name is the official one, and the standard one in the United States, for the dog still popularly known as the **Alsatian**. The breed in its present form came into existence in Germany as a particular crossbreed of different European shepherd dogs (i.e., sheepdogs).

**golden retriever** The name is that of a particular breed of retriever having a golden-red coat. The dog itself has been popular in Britain since just before World War I, when the Golden Retriever Club was founded.

**Gordon setter** The breed of setter was originally promoted by Alexander Gordon, 4th Duke of Gordon (1743–1827), and is named for him. The name was officially accepted for the breed in 1924. An alternate name is "black and tan setter," for the dog's coal-black coat with tan (chestnut) markings.

**great Dane** The dog of this breed originally hunted wild boar in Germany, then subsequently in Denmark and France. At first it was known simply as a "Dane," from the country where it hunted, but "great" was later added to distinguish it from the "lesser Dane," otherwise **Dalmatian**, which was (falsely) believed to be related to it. The modern name is first recorded in 1750 in the writings of the French naturalist Buffon, who refers to the dog as *le grand danois*.

greyhound (or grayhound)  The breed is a very old one, familiar to the Greeks and Romans, and long popular in coursing (hare-hunting). Its name does not refer to its color, but has evolved from Old English *grīghund*, the first half of which probably meant simply "dog," and was itself a word of Norse origin.

griffon  The name is particularly associated with the "griffon Bruxellois," or "Brussels griffon," which as its name implies originated from Brussels, Belgium. The word is itself French for "griffin," the mythical creature with the head of a bird and the body of a lion. The dog hardly resembles this, however, so was presumably so named for its unusual monkey-like head.

Groenendal  The dog, also known as "Belgian shepherd dog," takes its name from the small town of Groenendal, near Brussels, Belgium, where the breed was developed in the 19th century.

hackney  The breed of harness horse, with its distinctive high-stepping trot, has a name that may ultimately derive from Hackney, now a northeastern district of London, where it was originally pastured. Its name has become popularly confused with "hack," although this word properly denotes a refined riding horse. (Hence the "hacking jacket" as a term for a riding jacket.)

Himalayan  The name is that adopted in the United States for the cat known in Britain as a colorpoint, that is, a cat that is of Persian type but with the coat pattern (and eye color) of the Siamese. The name derives from that of the Himalayan rabbit, whose coat has the same coloration.

Holstein  The German breed of horse, used for both riding and driving, takes its name from the region of northern Germany where it was pastured on the right bank of the Elbe.

hunter  The name is not strictly speaking that of a breed but of a horse used for hunting, whether the quarry is fox, hare, boar, or stag. If a particular breed is to be associated with the chase, then it would almost certainly be the thoroughbred.

husky  The "Eskimo dog" or "sled dog" has a name that might seem to refer to its "husky" voice, or to the fact that it is big and strong and so "husky." In fact the name is simply a shortened form of *Eskimo*, so indicates the origin of the breed as a sledge dog of the Arctic, where it is still used by Eskimos as a reliable form of transport.

Irish setter  The name refers to the variety of setter that originated in

Ireland, as distinct from the English setter or any other kind. It is also known as the "red setter," for the glowing chestnut color of its coat.

**Jack Russell** The terrier is named for its original breeder, the so called "sporting parson" of Devon, England, the Rev. John Russell (1795–1883), who evolved it from the fox terrier with the aim of having a dog that would both run with the hounds and bolt (flush or dig out) the foxes on his native Exmoor. The original strain that he bred is now extinct, so that the present dogs of the name are of a different variety.

**keeshond** Originally known as the "Dutch barge dog," from its traditional work in guarding barges on Dutch canals, the breed has a name that was officially approved by the Kennel Club in 1926. It is said to derive from the Dutch personal name *Kies,* a short form of *Cornelis,* "Cornelius," and Dutch *hond,* "dog." It is not known who the named individual was, or even whether he was a human or another dog.

**kelpie** Also known as the "Australian sheepdog," the breed evolved as a blend of Scottish collie and dingo in about 1870. It is named for one of the original dogs so bred, itself probably named for the *kelpie,* or Scottish water sprite.

**Kerry blue** The breed is the largest of the Irish terriers that are properly native to Ireland. It is named for County Kerry, where it was originally popularized before being introduced to England in the 1920s. The second word of the name refers to the color of its bluish-gray coat.

**King Charles spaniel** The breed of spaniel is named for the English king Charles II, who popularized it in the 17th century, together with his brother James. Its preferred American name is "English toy spaniel." It exists in four varieties: (1) the black and tan, or King Charles proper, so named for its glossy black coat with tan (chestnut) markings; (2) the tricolor, for its white coat with black patches and tan markings; (3) the ruby, for its rich chestnut-colored coat; (4) the Blenheim, having a white coat with chestnut-red patches, popularized at Blenheim Palace, Oxfordshire, seat of the Duke of Marlborough. See also **Cavalier King Charles spaniel.**

**Korat** The shorthair cat of this name is a breed imported some time ago from Thailand to the United States, and more recently to Britain. It takes its name from the province in southern Thailand where it originated.

**kuvasz** The ancient breed of Hungarian guard dog and hunting dog, popular with the nobility in the 15th century, has a name that simply means "guard," from Turkish *kavas.* The dog is today used as a sheep and cattle drover, and although introduced to the United States, remains virtually unknown in Britain.

**Labrador** The familiar breed of retriever takes its name from the Canadian province where it was developed as a water dog in the 19th century. The dog became popular with English hunters, and was imported by them to England on board fishing vessels returning from Labrador.

**laika** The Russian breed of dog, or more exactly group of breeds, originated as a hunting dog in northwestern Russia. Its name derives from Russian *lay* (pronounced "lie"), meaning "bark," since when the dog sees a bird or animal in a tree it barks to attract the creature's attention and continues barking until the hunter arrives. Its function is thus the equivalent of that of a **pointer.**

**Landseer** The dog is a type of Newfoundland, black and white in color, and is named for the noted English animal painter Sir Edwin Landseer, who portrayed it in several of his paintings of the 1820s. As a result of this, many people wrongly assumed that it was the only type of true Newfoundland.

**Lhasa apso** The breed of Tibetan terrier was imported to England in the early 20th century, taking its name from Lhasa, the Tibetan capital, with *apso,* the Tibetan word for "terrier."

**Lipizzaner** (or **Lippizaner**) The famous breed of white horse takes its name from Lipizza, now in northwestern Croatia, near Trieste, where it evolved at the Austrian Imperial Stud, itself founded in 1580 by the Archduke Charles, son of the Austrian emperor Ferdinand I. The horse is now associated with the Spanish Riding School at Vienna, where it is taught the difficult movements of *haute école* ("high school").

**Maine coon** As its name indicates, the breed of cat originated in the New England state of Maine, where it evolved as a cross between domestic shorthair cats and imported longhairs. It is now found also in Britain. The second word of the name appears to imply that the animal was originally a hybrid between a cat and a raccoon, or between a cat and some other member of the cat family, such as a lynx.

**Malamute** The breed of dog, also known as "Alaskan Malamute," resembles a husky, although is distinct from it. It takes its name from the Eskimo tribe on the Alaskan coast for whom it worked as a sledge dog.

**Manipur** The breed of pony is named for the state in northeastern India where it was trained for use in polo, a game introduced in the 7th century by the reigning king of Manipur.

**Manx** The famous tailless breed of cat takes its name from the Isle of Man. The origin of the breed is uncertain, both with regard to its ultimate place of origin and its missing tail. The term "Manx cat" is sometimes used for any tailless cat, although the Manx is really a distinctive breed in its own right.

**mastiff** The breed is centuries-old, dating back to pre-Christian times, and was well known as a fighting dog in Asia and Africa. In Europe it was known to the Romans, while in England it was used for bear-baiting. Its name ultimately derives from Latin *mansuetus,* "tamed," literally "trained by hand," denoting a dog that, although strong and powerful, had been successfully domesticated by man.

**Morgan** The American breed of small, compact saddle horse is named for its original progenitor, the horse JUSTIN MORGAN, foaled in 1789 in the Green Mountain country of Vermont. The horse, a bay originally named FIGURE, was itself named for its owner, Thomas Justin Morgan (1747–1798). The horse died in 1821, having established one of America's most famous general purpose breeds.

**mustang** The small breed of horse was originally associated with the semi-wild horses of the plains of western America, especially those of California and Mexico. Its name represents Mexican Spanish *mestengo,* "strayer," from *mesta,* a term for an association of graziers, one of whose functions was to appropriate wild cattle that had become attached to the tame herds. The wildest type of mustang, and the one that was hardest to tame, came to be known as a "bronco," from Mexican Spanish *potro bronco,* "unbroken colt." Hence "bronco buster" as a term for a cowboy who breaks in broncos or wild horses.

**New Forest** The familiar English breed of pony is the country's largest. It takes its name from the extensive area of woodland in Hampshire, southern England, which was created (and expanded) by William the Conqueror in the 11th century as a hunting preserve. The ponies still roam wild there today in what is more a sparse wasteland than a full-blown forest.

**Newfoundland** The "Newfie," to give the breed its popular short name, is related to the **St. Bernard,** and is obviously named for the Canadian island and province. It is still uncertain, however, how the dog came to Newfoundland in the first place. One theory claims that it was brought there in the 17th century by Basque fishermen, but documentary evidence for this is unreliable.

**Old English sheepdog** The name of the well-known "woolly bear" is self-explanatory, although today the dog is rarely used as a sheepdog. It is also known as the "bobtail," for its bobbed (i.e., docked) tail. Many Old English sheepdogs are traditionally named BOB with reference to this feature.

**Orlov** The famous breed of Russian trotting horse is named for its originator, Count Aleksey Grigorievich Orlov (1737–1808). The Count, together with his brother Grigory, is historically notorious for leading the coup that resulted in the death of Czar Peter III and that brought Catherine, Grigory's mistress, to the throne.

**otterhound**   The breed of this name today is of a purer strain than the otterhounds that hunted otters for King John in the 13th century. The dog itself resembles a bloodhound, and is obviously a water dog.

**palomino**   The strikingly beautiful horse of North America, the "golden horse of the West," with its cream or white mane and tail, is not strictly a breed. The name thus refers to its color, and is American Spanish in origin, from *palomino,* a diminutive of *paloma,* "dove." Certain ringdoves have coloring similar to that of the horse. The name does not thus derive, as sometimes explained, from one Juan de Palomino, to whom Cortés, the Spanish conqueror of Mexico, is said to have presented such a horse in the 16th century. It is on record, however, that Cortés had the horses in Mexico in 1519.

**papillon**   The name of the breed is the French word for "butterfly," referring to the small dog's ears, which stand erect like butterfly wings. In the 16th century the breed was known as the "dwarf spaniel."

**Pekinese** (or **Pekingese**)   Popularly known as the "Peke," the small breed of dog takes its name from the Chinese capital. It was introduced to Britain in 1860 when the Summer Palace in Peking was sacked and Allied officers brought back some of the Imperial pets, presenting one, named LOOTIE (see Chapter 10, page 122), to Queen Victoria.

**Percheron**   The heavy breed of draft horse, familiar today outside its native France, takes its name from the ancient region of Perche, in northwestern France, where it originated.

**Persian**   The well-known variety (not breed) of cat has a name that is now virtually synonymous with "longhair," so that all longhairs can be popularly called "Persian." Originally, there were two distinctive varieties of longhair cat: the **Angora** (see above) and the Persian, the former from Turkey, the latter from Persia (modern Iran). However, the two types were interbred in an indiscriminate manner, so that the distinctive characteristics of each were blended or lost. Some of the original Persians used for pedigree breeding in the last quarter of the 19th century were said to have come from the palace of the Shah himself.

**Pharaoh hound**   The rather impressive breed of dog was introduced to Britain from Malta in the mid–20th century, and is so named for its resemblance to the dogs of the Pharoahs seen in ancient paintings of Egyptian hounds. Its own ancestry probably ultimately goes back to such hounds, which would themselves have been taken to Malta by the Phoenicians.

**pinto**   The familiar "painted horse" of America, well known to readers of

Wild West stories, derives its name from the American Spanish word that actually means "painted," referring to the horse's striking black and white markings. In Britain horses of this type are generally known as "piebald" or "skewbald."

**pointer** The breed originated in Spain, like the spaniel, despite its English name. It is so called since as a hunting dog it "points" when it detects a gamebird by standing still with its nose, body, and tail in a straight line. It has adopted a new role in the United States, where it now retrieves game. Compare the name of the **setter**, which indicates the presence of game in a different manner.

**Pomeranian** The miniature breed, related to the keeshond and Samoyed, was originally larger than the toy dog of today. It was introduced to Britain from the region of Pomerania, north-central Europe (now chiefly in Poland, bordering the Baltic). The "Pom" is known in French as a *loulou*, a term of affection corresponding to English *lulu*.

**poodle** In continental Europe, the breed was originally a water dog, and this explains the name, which represents German *Pudel*, a short form of *Pudelhund*, from *pudeln*, "to puddle about," "to splash in water." The dog has become associated with the French, who transformed it into a lap dog. The French name for the breed, *caniche*, refers to its original function, however, since it derives from *canard*, "duck," this bird being the one that the dog was trained to retrieve from the water. See also the quoted passage under BOY, the name of Prince Rupert's poodle (Chapter 10, page 108).

**Przewalski's horse** The name is that of the rare wild horse of western Mongolia that was named for the Russian explorer who discovered it in 1876, Nikolai Mikhailovich Przewalski (1839–1888). The horse was officially named *Equus przewalskii* in his honor in 1881 by I.S. Polyakov of the Russian Geographical Society.

**quarter horse** The small, powerful horse derives its name from the quarter-mile racetracks in Virginia for which it was bred in the late 18th century. Genealogically the breed stems from the English thoroughbred JANUS, who raced in Virginia and North Carolina between 1756 and 1780.

**retriever** The dog is so named since it was originally trained to set up (start up) game again when it had been lost, that is, to find it again (compare modern French *retrouver* in this sense). Today, however, it actually retrieves or recovers game. The sense of "retrieve" generally in English has similarly altered, meaning originally "find again" but now "recover," "get back."

**Rottweiler** The breed of dog, formerly used for boar-hunting in continental

Europe, takes its name from its place of origin, the town of Rottweil in Baden-Württemberg, western Germany.

**saddle horse**  The breed of horse has its origin in the early pioneering days of the first settlement in America in 1565, when there were only two means of movement over the vast distances of the new continent. One was by water, and the other on horseback. There was no indigenous horse in America, so the early settlers brought their own and bred them as light, strong, and speedy saddle horses able to cover the long distances.

**St. Bernard**  For the origin of the breed name, see the individual dog name BARRY in Chapter 10, page 106.

**Saluki**  The breed of dog is an ancient type of greyhound, popular as a working dog in Arab countries. It was introduced to England in the 19th century, and takes its name from Saluq, an ancient city of southern Arabia, where it developed.

**Samoyed**  The dog takes its name from the Samoyeds, a tribe of Siberian Mongol people, who used it as a sledge dog. The tribe itself is popularly said to derive its name from a Russian word meaning "self-eater" (just as *samovar* means literally "self-boiler"), referring to cannibalistic practices. But this origin is now discounted, and the name means simply "land of the Saamians," referring to a Lappish race which emigrated to this part of Russia. Some Samoyed dogs are given the generic name of **laika** (see above).

**schipperke**  Just as the keeshond is also known as a "Dutch barge dog," so the schipperke has the alternate name of "Belgian barge dog." Like the keeshond, it had the original role of guarding barges and river boats on Belgian waters. Its name reflects such work, and represents the Flemish word for "little skipper." The breed was introduced to Britain in the late 19th century.

**schnauzer**  The dog is a type of terrier introduced to Britain and the United States in the 1920s and 1930s from German-speaking countries. German *Schnauze* means "snout," although the dog's muzzle is not particularly distinctive, nor is it even a "sniffer dog."

**Scottish terrier**  The popular "Scottie" is now an individually recognized breed of dog, although originally the name was fairly general for any kind of Scottish terrier, such as the cairn, Skye, West Highland white, or even the Dandie Dinmont. Its name is of obvious origin, and individual dogs are still given traditional Scottish names such as JOCK and MAC.

**Sealyham**  The terrier of this name was bred in 1851 for badger-baiting by Captain John Tucker-Edwardes of Sealy Ham House, south of Fishguard,

Dyfed, southwestern Wales, and so was named for the house (now itself known as Sealyham Hall or simply Sealyham).

**setter** The name was originally used of the spaniel, although there are now three distinct types of setter: the **English setter, Irish setter,** and **Gordon setter.** The word itself refers to the dog's ability to "set," that is, to crouch down at or near the point where it discovers hunted game. Compare the name of the **pointer,** which indicates the presence of game in a different manner.

**sheepdog** The self-descriptive name is used of a number of breeds who work to herd or drive sheep. Among the best known are the collie and the Old English sheepdog, although the latter now rarely works with sheep. See also the name of the **Shetland sheepdog.**

**Shetland pony** The pony, one of the smallest breeds, is familiar in many countries of the world. It is named for the Shetland Islands, northern Scotland, where it is known to have existed many centuries ago. Its exact place of origin, however, is uncertain. Like the Shetland sheepdog (see below) it is colloquially known as a "Sheltie."

**Shetland sheepdog** The small breed works to herd or drive small sheep in the Shetland Islands, northern Scotland, from which it takes its name. The breed resembles a small collie, and the name originally proposed for the dog was actually "Shetland collie." Collie owners objected, however, and the present name was instead officially adopted in 1914. The dog's nickname of "Sheltie" is also used for the **Shetland pony** (see above).

**shih tzu** The breed of dog was developed from the Lhasa apsos which were offered as tributes to the emperors of China. The Chinese name means "lion dog" (in full, *shih-tzu kou,* literally "lion son dog"), from its fancied resemblance to this animal, either physically or symbolically. The breed was introduced to England in the 1930s.

**shire horse** The large and heavy breed of carthorse is named for "the Shires," that is, the Midland (central) counties of England, such as Leicestershire and Northamptonshire, where it was long used as a draft horse in agricultural work. The breed is colloquially known simply as the "shire."

**Siamese** The well-known and popular breed of cat has a head and body that resemble those of the ancient cats once worshipped in Egypt. The breed name, however, refers to the more immediate origin of the cats, as the first of the type were brought to England in 1884 from the Royal Palace in Bangkok, capital of Siam (now Thailand). It is thus held that the breed as it exists today evolved from the historic royal cats of Siam.

**Skye terrier** The breed of terrier was originally used for bolting (flushing out) badgers and foxes on the Scottish island of Skye. Hence its name.

**Somali**  The breed of cat, a marked Abyssinian type, is a relatively recent one. It was given a name on a geographical basis to indicate this particular affinity, as Somalia in northeastern Africa is a country bordering on Ethiopia, formerly Abyssinia.

**spaniel**  The breed is a fairly old one, and is mentioned in Chaucer (14th century). The name indicates the dog's country of origin, since it derives from Old French *espaigneul* (modern *épagneul*), itself from *Espagne,* "Spain." Despite this particular provenance, all the main breeds of spaniel today are English, such as the Clumber, cocker, and springer.

**springer spaniel**  Officially known as the "English springer spaniel," the breed is so called since the dog was trained to "spring" game, that is, to make the gamebirds fly up suddenly so that (originally) they were caught in a net or pounced on by a falcon. In essence, therefore, the function of a springer was the same as that of a cocker. Today the springer has a more general use as a sporting dog for finding and retrieving game, as well as flushing it up. It was at one time known as the "Norfolk spaniel," from its association with the estates of the Duke of Norfolk.

**Staffordshire bull terrier**  The dog is a specialized breed of bull terrier, originating, as its name indicates, in the county of Staffordshire, central England. It reached southern England only in the 1920s, although it was familiar in the United States from as early as the 1870s.

**staghound**  The dog's name indicates its former use for hunting stags. In Norman times it worked in particular in the New Forest (see **New Forest pony,** above).

**standardbred**  The North American breed of trotting and pacing horse is well known from devotees of harness-racing. It is so named as it conforms to the precise standards of breeding laid down in 1879 by the National Association of Trotting Horse Breeders. The name is also intended to distinguish this horse from the **thoroughbred,** from which it differs by its heavier limbs and more robust build. The ancestor of the modern standard trotter is HAMBLETONIAN (see Chapter 10, page 118), foaled in 1849.

**Suffolk punch**  The well-known breed of English draft horse, always chestnut in color, derives the first part of its name from the county of eastern England where it originated in the early 16th century. The second part of the name represents a dialect word for a short and thickset animal.

**Sussex spaniel**  The strain of spaniel was first bred in the county of Sussex, southern England, in the late 18th century, with the pioneer breeder a Mr. Fuller of Rosehill Park, near Hastings. His dogs were the ancestors of most of today's breed.

**Sydney silky** The breed of terrier originated in and around Sydney, Australia, and was a descendant of the Australian terrier and the Yorkshire terrier (and also possibly the Skye terrier). The dog is not common in Britain, but is known in the United States and, of course, in Australia. The second word of the name refers to the dog's silky coat texture.

**tabby** The familiar domestic cat so called, with its distinctive striped coat and characteristic coat pattern, takes its name from the fabric known as "tabby" that its coloring and markings resemble. The fabric itself, a type of silk or taffeta with a wavy finish, is named for Al-'Attabiya, the quarter of Baghdad where it was originally made. At some stage the cat's name became associated with the pet form of the personal name Tabitha (or possibly the other way round), which is why "tabby" became a colloquial word for a gossiping or "catty" old woman, or in Australian English, a slang term for any girl or woman.

**Tennessee walking horse** The American breed of horse is sometimes called a "Plantation walking horse" or "Plantation walker." Like the **Morgan** (see above), the breed owes its origin to a single progenitor, in this case the stallion BLACK ALLAN (named for his color), foaled in 1886 and taken to Tennessee as a colt. The horse's popular name derives from its use to carry farmers and planters of the South over their plantations. A "walking horse" is one that moves at a fast and easy running walk, as distinct from a true running gait.

**terrier** The word, now applied to several individual breeds, derives from French *terre*, "earth," since the dog originally pursued its quarry, typically a fox or badger, into the animal's earth, from which it either dug it out or drove it out.

**thoroughbred** The name is that of the most familiar breed of horse, and is virtually synonymous with the modern racehorse. A full name used in Britain, "English thoroughbred," wrongly suggests that the horse originated in England. It is called "English" because it has been bred and developed for many years in England, but "thoroughbred" is a literal translation of its Arabic name, *kehilan*, which itself indicates its Arab origin. All thoroughbreds trace their ancestry to three Arab sires: the DARLEY ARABIAN (see page 113), the GODOLPHIN ARABIAN (see page 117), and the BYERLY TURK (see page 110). The descendants of these three horses were imported to England from about 1690 to about 1730.

**Turkish Van** A number of breeds of cat have originated in Turkey, the first being the Angora. The Turkish Van was introduced to Britain in 1956 and takes its name for the region surrounding Lake Van, in eastern Turkey, where it was first found.

**Weimaraner** The breed of dog was introduced to Britain only in the

1950s, although it has been known in the United States at least from the 1920s. It is a sporting dog of German origin, and was popular at the court of Weimar in the 19th century. Hence its name.

**Welsh cob**   The hardy breed of pony evolved from the Welsh mountain pony, still found living in a wild state in the Welsh mountains and wastes. The second word of its name is a general term for a thickset riding or draft horse.

**Welsh springer spaniel**   The breed of springer spaniel comes from Wales, as its name indicates, and in its land and language of origin is also known as the "targi," literally "dispersing dog," from Welsh *tarfu*, "to scare," "to scatter," and *gi*, a mutated form of *ci*, "dog" (compare the name of the **corgi**). The name refers to its role in springing game.

**West Highland white terrier**   The lengthy breed name, colloquially abbreviated to "Westie," relates to the dog's origin in the West Highlands of Scotland. It is "white" because it is the sole all-white Scottish breed.

**whippet**   The present dog of the name probably had the greyhound and a mixture of terriers as its ancestors. The name probably refers to its ability to "whip" along swiftly when running, or to "whip off" by moving quickly when starting to run.

**Yorkshire terrier**   The breed evolved in the 19th century in the West Riding of Yorkshire, northern England. The "Yorkie" was at first known as a "broken-haired Scottish terrier," referring to the way in which its coat is "broken," or parted on its back from nose to tail so that it is long enough to sweep the ground.

# APPENDIX II: HOUND NAMES

There are many names that have long been traditional for hounds (see Chapter 8). One of the most comprehensive listings of such names is that of over 7,000 in Daphne Moore's *The Book of the Foxhound* (see Bibliography). However, this is rather too generous for our purposes, so here is the more modest tally of 803 traditional names listed in Thelma Gray, *The Beagle,* 4th ed. (London: Popular Dogs Publishing, 1980):

## Dogs

| | | | |
|---|---|---|---|
| Actor | Banger | Capitol | Climbank |
| Adamant | Barbarous | Captain | Clinker |
| Adjutant | Bellman | Captor | Combat |
| Agent | Bender | Carol | Combatant |
| Aider | Blaster | Carver | Comforter |
| Aimwell | Bluecap | Caster | Comrade |
| Amorous | Blueman | Castwell | Comus |
| Antic | Bluster | Catcher | Conflict |
| Anxious | Boaster | Catchpole | Conqueror |
| Arbiter | Boisterous | Caviller | Conquest |
| Archer | Bonnyface | Cerberus | Constant |
| Ardent | Bouncer | Challenger | Contest |
| Ardor | Bowler | Champion | Coroner |
| Arrogant | Bragger | Charon | Cottager |
| Artful | Bravo | Chaser | Counsellor |
| Artist | Brawler | Chaunter | Countryman |
| Atlas | Brazen | Chieftain | Courteous |
| Atom | Brusher | Chimer | Coxcomb |
| Auditor | Brutal | Chirper | Craftsman |
| Augur | Burster | Claimant | Crasher |
| Awful | Bustler | Clamorous | Critic |
| Bachelor | Caitiff | Clangour | Critical |
| Baffler | Caliban | Clasher | Crowner |

| | | | |
|---|---|---|---|
| Cruiser | Forward | Jolly | Mittimus |
| Crusty | Fugleman | Jollyboy | Monarch |
| Cryer | Fulminant | Jostler | Monitor |
| Curfew | Furrier | Jovial | Motley |
| Currier | Gainer | Jubal | Mounter |
| Damper | Gallant | Judgement | Mover |
| Danger | Galliard | Jumper | Mungo |
| Dangerous | Galloper | Labourer | Musical |
| Dapper | Gamboy (sic) | Larum | Mutinous |
| Dapster | Gamester | Lasher | Mutterer |
| Darter | Garrulous | Laster | Myrmidon |
| Dasher | Gazer | Launcher | Nestor |
| Dashwood | General | Leader | Nettler |
| Daunter | Genius | Leveller | Newsman |
| Dexterous | Giant | Liberal | Nimrod |
| Disputant | Gimcrack | Libertine | Noble |
| Downright | Glancer | Lictor | Nonsuch |
| Dragon | Glider | Lifter | Novel |
| Dreadnought | Glorious | Lightfoot | Paean |
| Driver | Goblin | Linguist | Pageant |
| Duster | Governor | Listner | Paragon |
| Eager | Grapler (sic) | Lounger | Paramount |
| Effort | Grasper | Lucifer | Partner |
| Elegant | Griper (sic) | Lunger | Partyman |
| Eminent | Growler | Lurker | Pealer |
| Envious | Grumbler | Lusty | Penetrant |
| Envoy | Guardian | Manager | Perfect |
| Ernest | Guider | Manful | Perilous |
| Errant | Guiler | Marksman | Pertinent |
| Excellent | Hannibal | Marplot | Petulant |
| Factious | Harasser | Marschal | Phoebus |
| Factor | Harbinger | Martial | Piercer |
| Fatal | Hardiman | Marvellous | Pilgrim |
| Fearnought | Hardy | Matchem | Pillager |
| Ferryman | Harlequin | Maxim | Pilot |
| Fervent | Havoc | Maximus | Pincher |
| Finder | Hazard | Meanwell | Piper |
| Firebrand | Headstrong | Medler (sic) | Playful |
| Flagrant | Hearty | Menacer | Plodder |
| Flasher | Hector | Mendall | Plunder |
| Fleece'm | Heedful | Mender | Politic |
| Flinger | Hercules | Mentor | Potent |
| Flippant | Hero | Mercury | Prater |
| Flourisher | Highflyer | Merlin | Prattler |
| Flyer | Hopeful | Merryboy | Premier |
| Foamer | Hotspur | Merryman | President |
| Foiler | Humbler | Messmate | Presto |
| Foreman | Impetus | Mighty | Prevalent |
| Foremost | Jerker | Militant | Primate |
| Foresight | Jingler | Minikin | Principal |
| Forester | Jockey | Miscreant | Prodigal |

## Appendix II. Hound Names

| | | | |
|---|---|---|---|
| Prompter | Salient | Stroker | Trueman |
| Prophet | Sampler | Stroller | Trusty |
| Prosper | Sampson | Struggler | Trywell |
| Prosperous | Sanction | Sturdy | Tuner |
| Prowler | Sapient | Subtile (*sic*) | Turbulent |
| Pryer | Saucebox | Succour | Twanger |
| Racer | Saunter | Suppler | Twig'em |
| Rager | Scalper | Surly | Tyrant |
| Rallywood | Scamper | Swaggerer | Vagabond |
| Rambler | Schemer | Sylvan | Vagrant |
| Rampant | Scourer | Tackler | Valiant |
| Rancour | Scrambler | Talisman | Valid |
| Random | Screamer | Tamer | Valorous |
| Ranger | Screecher | Tangent | Valour |
| Ransack | Scuffler | Tartar | Vaulter |
| Rantaway | Searcher | Tattler | Vaunter |
| Ranter | Settler | Taunter | Venture |
| Rapper | Sharper | Teaser | Venturer |
| Ratler (*sic*) | Shifter | Terror | Venturous |
| Ravager | Signal | Thrasher | Vermin |
| Ravenous | Singwell | Threatner (*sic*) | Vexer |
| Ravisher | Skirmish | Thumper | Victor |
| Reacher | Smoker | Thunderer | Vigilant |
| Reasoner | Social | Thwacker | Vigorous |
| Rector | Solomon | Thwarter | Vigour |
| Regent | Solon | Tickler | Villager |
| Render | Songster | Tomboy | Viper |
| Resonant | Sonorous | Topmost | Volant |
| Restive | Soundwell | Topper | Voucher |
| Reveller | Spanker | Torment | Wanderer |
| Rifler | Special | Torrent | Warbler |
| Rigid | Specimen | Torturer | Warhoop |
| Rigour | Speedwell | Tosser | Warning |
| Ringwood | Spinner | Touchstone | Warrior |
| Rioter | Splendour | Tracer | Wayward |
| Risky | Splenetic | Tragic | Wellbred |
| Rockwood | Spoiler | Trampler | Whipster |
| Romper | Spokesman | Transit | Whynot |
| Rouser | Sportsman | Transport | Wildair |
| Router | Squabbler | Traveller | Wildman |
| Rover | Squeaker | Trial | Wilful |
| Rudesby | Statesman | Trier | Wisdom |
| Ruffian | Steady | Trimbush | Woodman |
| Rumbler | Stickler | Trimmer | Worker |
| Rummager | Stinger | Triumph | Workman |
| Rumour | Stormer | Trojan | Worthy |
| Runner | Stranger | Trouncer | Wrangler |
| Rural | Stripling | Truant | Wrestler |
| Rusher | Striver | Trudger | |
| Rustic | Strivewell | Trueboy | |

## Bitches

| | | | |
|---|---|---|---|
| Accurate | Concord | Fretful | Likely |
| Active | Courtesy | Friendly | Lissom |
| Actress | Crafty | Frisky | Litigate |
| Affable | Crazy | Frolic | Lively |
| Agile | Credible | Frolicsome | Lofty |
| Airy | Credulous | Funnylass | Lovely |
| Amity | Croney (*sic*) | Furious | Luckylass |
| Angry | Cruel | Fury | Madcap |
| Animate | Curious | Gaiety | Madrigal |
| Artifice | Dainty | Gaily | Magic |
| Audible | Daphne | Gainful | Matchless |
| Baneful | Darling | Galley | Melody |
| Bashful | Dashaway | Gambol | Merriment |
| Bauble | Dauntless | Gamesome | Merrylass |
| Beauteous | Delicate | Gamestress | Mindful |
| Beauty | Desperate | Gaylass | Minion |
| Beldam | Destiny | Giddy | Miriam |
| Bellmaid | Dian (*sic*) | Gladness | Mischief |
| Blameless | Diligent | Gladsome | Modish |
| Blithesome | Docile | Governess | Monody |
| Blowzy | Document | Graceful | Music |
| Bluebell | Doubtful | Graceless | Narrative |
| Bluemaid | Doubtless | Gracious | Neatness |
| Bonny | Dreadless | Grateful | Needful |
| Bonnylass | Dulcet | Gravity | Nicety |
| Boundlass (*sic*) | Easy | Guilesome | Nimble |
| Bravery | Echo | Guiltless | Noisy |
| Brevity | Ecstasy | Guilty | Notable |
| Brimstone | Endless | Handsome | Notice |
| Busy | Energy | Harmony | Notion |
| Buxom | Enmity | Hasty | Novelty |
| Capable | Fairmaid | Hazardous | Novice |
| Captious | Fairplay | Heedless | Passion |
| Careful | Faithful | Helen | Pastime |
| Careless | Famous | Heroine | Patience |
| Carnage | Fanciful | Honesty | Phoenix |
| Caution | Fashion | Hostile | Phrensy (*sic*) |
| Cautious | Favourite | Industry | Placid |
| Charmer | Fearless | Jealousy | Playful |
| Chauntress | Festive | Jollity | Playsome |
| Cheerful | Fickle | Joyful | Pleasant |
| Cherriper (*sic*) | Fidget | Joyous | Pliant |
| Chorus | Fiery | Laudable | Positive |
| Circe | Fireaway | Lavish | Precious |
| Clarinet | Firetail | Lawless | Prettylass |
| Clio | Flighty | Levity | Previous |
| Comeley (*sic*) | Flourish | Liberty | Priestess |
| Comfort | Flurry | Lightning | Probity |
| Comical | Forcible | Lightsome | Prudence |

## Appendix II. Hound Names

| | | | |
|---|---|---|---|
| Racket | Skilful | Tractable | Vitiate |
| Rally | Songstress | Tragedy | Vivid |
| Rampish (*sic*) | Specious | Trespass | Vixen |
| Rantipole | Speedy | Trifle | Vocal |
| Rapid | Spiteful | Trivial | Volatile |
| Rapine | Sportful | Trollop | Voluble |
| Rapture | Sportive | Troublesome | Waggery |
| Rarity | Sportly | Truelass | Waggish |
| Rashness | Sprightly | Truemaid | Wagtail |
| Rattle | Stately | Tunable | Wanton |
| Ravish | Stoutness | Tuneful | Warfare |
| Reptile | Strenuous | Vanquish | Warlike |
| Resolute | Strumpet | Vehemence | Waspish |
| Restless | Surety | Vehement | Wasteful |
| Rhapsody | Sybil | Vengeance | Watchful |
| Riddance | Symphony | Vengeful | Welcome |
| Riot | Tattle | Venomous | Welldone |
| Rival | Telltale | Venturesome | Whimsy |
| Roguish | Tempest | Venus | Whirligig |
| Ruin | Termagant | Verify | Wildfire |
| Rummage | Terminate | Verity | Willing |
| Ruthless | Terrible | Victory | Wishful |
| Sanguine | Testy | Victrix | Wonderful |
| Sappho | Thankful | Vigilance | Worry |
| Science | Thoughtful | Violent | Wrathful |
| Scrupulous | Tidings | Viperous | Wreakful |
| Shrewdness | Toilsome | Virulent | |

# APPENDIX III: LITERARY LISTINGS

Many fictional works, from the Greek and Roman classics to modern novels, are remarkable for their animal characters. The sections below list the animals found in the writings of seven authors.

Translations of foreign names are given wherever possible, as well as the origins of unusual or interesting names, where they are known. Some of the names, especially the better known ones, or those explained only briefly here, have their own individual entries (alphabetically listed) in Chapter 12, in which case their particulars below are followed by "(Chapter 12)."

## 1. The Hounds of Actaeon

In Greek mythology, the hunter Actaeon was changed into a stag by the goddess Artemis (Diana), whom he had come upon when she was bathing, and torn to pieces by his own hounds. They are traditionally 50 in number (25 couple, in modern hunting terminology), and Ovid, in his *Metamorphoses,* gives the names of 40 of them. Arranged alphabetically they are as follows, with translations:

AELLO, "whirlwind"
AGRE, "hunter"
AGRIODOS, "wild track"
ALCE, "powerful"
ARGUS, "bright," "swift" (see Chapter 11)
ASBOLUS, "sooty"
CANACHE, "barker"
DOÖRGA, "double angry"
DORCEUS, "leaper" (literally, "antelope")
DROMAS, "runner"
HARPALUS, "grasper"
HARPYEA, "snatcher"
HYLACTOR, "barker"
HYLEUS, "barker"
ICHNOBATES, "bramble track"
LABROS, "rowdy"
LACHNE, "shaggy"
LACON, "strong" (literally, "Laconian," "Lacedaemonian," "Spartan")

LADON, "embracer" (literally, "Ladon," the name of a 100-headed serpent)
LEBROS, "rowdy"
LELAPS, "storm" (see LAELAPS in Chapter 11)
LEUCITE, "white"
LEUCOS, "white"
LYCISA, "wolf"
MELAMPUS, "black foot"
MELANCHAETUS, "black hair"
MELANEUS, "inky"
MOLOSSUS, "Molossian" (a breed of wolfhound)
NAPE, "glen"
NEBROPHONOS, "deerslayer"
ORESTROPHOS, "mountain-bred"
ORIBASUS, "mountain ranger"
PACHYTOS, "thick-skinned"
PAMPHAGUS, "all-devouring," "omnivorous"
POEMENIS, "shepherd"
PTERELAS, "feather launcher"
STICTE, "spot"
THERIDAMAS, "beast tamer"
THERON, "beast"
THOUS, "eager"

## 2. Animals in Scott

The works of Sir Walter Scott (1771–1832) feature a large number of animals, mainly dogs and horses. In some cases the fictional animals are based on Scott's own. (For examples of the better known ones, see Chapter 10, page 105.) Below follows a listing of all animals included in M.F.A. Husband, *A Dictionary of the Characters in the Waverley Novels of Sir Walter Scott* and E. Thornton Cook, *Sir Walter's Dogs* (see both in Bibliography). Where a character was based on a real animal owned by Scott, the original is named in square brackets (as "modeled on").

APOLLYON, a jet black horse
BALDER, a "grisly old wolf-dog"
BALL, a palfrey (i.e., a ladies' riding horse)
BAN, a deerhound
BASH, a staghound
BATTIE, a staghound
BAWTY, a dog
BAYARD, (1) a palfrey; (2) a horse
BELZIE, a bulldog
BENEDICT, a palfrey
BERGEN, a palfrey
BERWICK, a charger
BEVIS, a "wolf dog in strength, a mastiff in form" [modeled on MAIDA] (Chapter 12)
BINGO, "a little mongrel cur"
BLACK HASTINGS, a warhorse
BLACK MOOR, a horse
BRAN, a greyhound
BUNGAY, a dog
BUSCAR, a greyhound
CHARLOT, a spaniel
CHEVIOT, a falcon
CROMBIE, a cow
CRUMMIE, a cow
DERMID, a mare
DIAMOND, a falcon
DOBBIN, a horse
DUMPLE, a pony
DUSTIEFOOT, a pony
ELPHIN, a "small cocking spaniel"
FAIRY, a Manx pony
FANGS, a "ragged, wolfish-looking dog"
GAUNTLET, a nag
GIBBON, a monkey
GORGON, a brache (i.e., a bitch hound used in hunting)
GOWANS, a "brockit cow" (i.e., a black and white one)
GREY GILBERT, a horse
GRIZZY, a cow

192                The Naming of Animals

GUSTAVUS, a horse
HEMP, a lurcher dog
JEMIMA, a mare
JEZABEL (sic), a Flemish mare
JUNO, a spaniel
KILLBUCK, a greyhound
KILSYTHE, a horse
LUCY, a spaniel
LUFRA, "the fleetest hound in all the North" (Chapter 12)
MAHOUND, an Arab horse
MOORKOFF, a horse
NEPTUNE, a dog
PESTLE and MORTAR, a pair of ponies
PHOEBE, a horse
PIXIE, a pony
PLATO, a spaniel
RANGER, a dog
ROAN ROBIN, a horse
ROSABELLE, a horse
ROSWAL, a staghound [modeled on MAIDA] (Chapter 12)
SATAN, a nag
SHAGRAM, an old pony
SNAP, a small dog
SOLOMON, a "useful iron-grey galloway" (a breed of small strong horse)
SORREL, a hunter
SOUPLE SAM and SOUPLE TAM, a pair of horses
STUMAH, a dog (Chapter 12)
TARRAS, a horse
TEAR'UM, a mastiff
THETIS, a dog
THRYME, a wolfhound
TRIMMER, a hound
WASP, a "rough terrier dog"
WHISTLER, a greyhound
WILDBLOOD, a horse
WOLF, a staghound
WOLF-FANGER, a bloodhound
YARROW, a sheepdog
YSEULTE, a palfrey
ZAMOR, a horse

Several of these names reflect the Scottish setting of the novels. The cows CROMBIE and CRUMMIE have names that may derive from Gaelic *crom* or Old English *crumb*, "crooked," for their "crumpled" horns. On the other hand, if the names relate to their shaggy coats, they could perhaps represent *crombie* as the term for a type of woollen cloth manufactured in Aberdeen. MAHOUND is the devilish name of the prophet *Muhammad*, more recently familiar (or notorious) from its occurrence in Salman Rushdie's novel *The Satanic Verses* (1988). WOLF-FANGER has a German name, meaning "wolf catcher." He comes in *Anne of Geierstein* (1829), largely set in Switzerland. YARROW was the namesake of a real dog, a sheep stealer, that Scott had read about.

## 3. Hounds in Surtees

The English novelist R.S. Surtees (1803–1864) satirized the sporting life of the gentry in such works as *Jorrocks's Jaunts and Jollities* (1838). His writings contain the names of many horses and hounds, the latter numbering almost 100, most of them in *Handley Cross* (1843). They include the following:

| | | |
|---|---|---|
| ADAMANT | DORIMONT | RALLYWOOD |
| AFFABLE | DREADNOUGHT | RANTER |
| BARBARA | DUSTER | RAVAGER |
| BARBICAN | FACTOR | RAVENOUS |
| BEAUFORT POTENTATE | FERRYMAN | RESOLUTE |
| BELMAID | FRANTIC | RUMMAGER |
| BENEDICT | FUGLEMAN | SINGLET |
| BILLINGSGATE | FUNNYGLASS | SOLOMON |
| BLAMELESS | FURRIER | SPLENDOUR |
| BLUEBELL | GALLOPER | STATESMAN |
| BOISTEROUS | GAMESTER | STRUMPET |
| BONIFACE | HARMONY | SULTAN |
| BONNETS-O'BLUE | HURRICANE | TAPSTER |
| BONNY-BELL | JACK | THUNDERER |
| BOUNCER | JINGLER | TIPLER (*sic*) |
| BRAVERY | JOLLY-BOY | TOM |
| BRILLIANT | JOUSTER | TOWLER |
| BRUSHER | JUDGEMENT | TRUEBOY |
| CHALLENGER | JUMPER | TRUSTY |
| CHEAPSIDE | LAVENDER | TUNEABLE |
| CHRISTIAN | LIGHTNING | VAGRANT |
| CLIMBANK | LIMNER | VANQUISHER |
| COME-BY-CHANCE | LOUISA | VENGEANCE |
| CONQUEROR | LUCIFER | VENUS |
| CONTEST | LUCKY-LASS | VICTOR |
| CORONER | MANAGER | VICTORY |
| COTTAGES | MARQUIS | VIGILANT |
| COUNTRYMAN | MERCURY | VILLAGER |
| CROWNER | MISCHIEF | WARBLER |
| CRUISER | MOUNTAIN | WARRIOR |
| DAUNTLESS | PILGRIM | WORKMAN |
| DEXTEROUS | PRIESTESS | |

It will be seen that many of these occur in the list of traditional names for hounds in Appendix II. For an example of a Surtees dog who is not a hound, see PONTO in Chapter 12, page 156. For Surtees horses, see below.

## 4. Horses in Surtees

The following names of horses occur in the writings of the English novelist R.S. Surtees (see above). Most of them are in *Handley Cross* (1843).

ARTERXERXES (*sic*), so called because he came "after Xerxes" when Mr. Jorrocks drove the gig-cart
BADAJOZ, from the city in southern Spain, retaken by the British from the French in 1812, with perhaps also a pun on "bad hoss"
BARNEY BODKIN

194                The Naming of Animals

BARROSA, from the place in southern Spain where the British defeated the French in 1811
BILLY BUTTON
BULL-DOG
BUSINESS
CLEVER CLUMSY, for a horse who is both; the earliest record of this phrase in the *Oxford English Dictionary* is 1854
CORUNNA, for the Spanish city (La Coruña), where the English defeated the French in 1809; perhaps also a pun on "runner"
DICKEY COBDEN, presumably for the British statesman *Richard Cobden* (1804–1865), who successfully campaigned to abolish the Corn Laws
DISMAL GEORDY
ECLIPSE, doubtless for the famous racehorse (see page 115)
FAIR ROSAMUND, from the nickname of *Rosamond* Clifford, mistress of Henry II in the 12th century, and the subject of various tales
FLORIZEL, for the character in Shakespeare's *The Winter's Tale*
FOXHUNTER, doubtless for a racehorse of this name
FOXHUNTORIBUS, a quasi–Latin pun on the preceding
GIMCRACK, probably for the famous racehorse (see page 117)
GINNUMS
HARKAWAY
HIGHFLYER
JUMPER
JUNIPER
ROSINANTE, for the horse of Don Quixote (see Chapter 12, page 159)
SALAMANCA, for the city in Spain where the British defeated the French in 1812
SKYLARK
SONTAG
STAR-GAZER
TALAVERA, for the Spanish city near which the British and Spanish defeated the French in 1809
TOULOUSE, for the town in southern France, where the British defeated the French in 1814 in the last battle of the Peninsular War
XERXES, see ARTERXERXES above
YOUNG HYSON

Many of these names were almost certainly given for well-known racehorses, who themselves may have been named for the peoples and places mentioned. Several of the names are those of battles in the Peninsular War (1808–14).

In *Mr. Facey Romford's Hounds* (1865), Surtees comments that horses bought from a dealer's yard were "generally christened with high-sounding names diametrically opposite to their respective qualities." Thus EVERLASTING slowed down and stopped on the slightest hill, HEARTY HARRY "wanted no end of codling," TWICE-A-WEEK "would hardly come out once in a fortnight," and GLUTTON "looked as if he had lived altogether upon toothpicks and water."

## 5. Animals in Dickens

The following animals occur in the works of Charles Dickens (1812–1870), and are taken from John Greaves, *Who's Who in Dickens* and Cumberland Clark, *The Dogs in Dickens* (see both in Bibliography).

*Appendix III. Literary Listings* 195

BOXER, a dog (Chapter 12)
BULL'S EYE, "a white shaggy dog" (Chapter 12)
CAPRICORN, a horse
CARLO, a performing dog, in the same troupe as PEDRO (below)
CAULIFLOWER, a horse
DAPH and JUNO, pointers (Chapter 12)
DICK, a pet blackbird
DIOGENES, a dog, with pet name DI (Chapter 12)
"EDDARD," a donkey
GRIP, a raven (Chapter 12)
JIP, a small spaniel (Chapter 12)
LADY GREY, a large gray cat
LION, a dog (Chapter 12)
LOVELY, a dog (Chapter 12)
MERRYLEGS, a performing dog (Chapter 12)
PEDRO, a performing dog, in the same troupe as CARLO
PINCHER and NEPTUNE, a pair of watchdogs
PONTO, a sporting dog (Chapter 12)
POODLES, a dog
WHISKER, a pony (Chapter 12)

As can be seen, many of these characters have their individual entries in Chapter 12. Among the most remarkable names in Dickens, however, are those of the 25 cagebirds belonging to the kindly but crazy old woman, Miss Flite, in *Bleak House* (1853). They are, in the order given: HOPE, JOY, YOUTH, PEACE, REST, LIFE, DUST, ASHES, WASTE, WANT, RUIN, DESPAIR, MADNESS, DEATH, CUNNING, FOLLY, WORDS, WIGS, RAGE, SHEEPSKIN, PLUNDER, PRECEDENT, JARGON, GAMMON, and SPINACH. Read consecutively, they tell a tale of progression from birth to death, youth to old age, sanity to insanity. Perhaps symbolically, the birds are all eventually released.

## 6. Animals in Dostoyevsky

A number of animals feature in the works of the famous Russian writer Fyodor Dostoyevsky (1821–1881). The names that follow are taken from Charles E. Passage, *Character Names in Dostoyevsky's Fiction* (see Bibliography). Most of the animals are dogs (no fewer than 15), although other kinds are also found. English translations are given for the names, which are mostly of Russian origin, but occasionally French.

AMISHKA, a dog, from French *ami*, "friend," with Russian diminutive suffix
AZORKA, a dog, for the hero of Grétry's opera *Zémire et Azor* (1771), with Russian diminutive suffix; compare ZEMIRKA, below
BYELKA, a dog, Russian for "squirrel"
GNEDKO, an old workhorse, and a young one replacing him, from Russian *gnedoy*, "sorrel"
HECTOR, a dog
KUL'TYAPKA, a dog, from Russian *kul'*, "sack," and *tyapka*, "cleaver"
MIMI, a lapdog, from the common French pet name
NORMA, a dog, for the heroine of the Bellini opera of 1831
PEREZVON, a dog, Russian for "peal of bells"; see ZHUCHKA, below

SHARIK, a dog, Russian for "little ball" (diminutive of *shar*)
SIR JOHN FALSTAFF, a dog
TANCRED, a dog, for the hero of Rossini's opera *Tancredi* (1815)
TRESORKA, a dog, from French *trésor*, "treasure," with Russian diminutive suffix
VAS'KA, a goat, from a diminutive of Russian *Vasiliy*, "Basil"
VOLOCHOK, a dog, from a diminutive form of either Russian *volk*, "wolf," or *volchok*, "spinning top"

ZEMIRKA, a dog, for the heroine of Grétry's opera *Zémire et Amor* (1771), with Russian diminutive suffix; compare AZORKA, above
ZHUCHKA, a dog, probably from French *joujou*, "toy," "plaything," with Russian diminutive suffix; not likely to be from Russian *zhuk*, "beetle," "bug"; the name is a common Russian one for a small dog
ZHUCHKA, another dog, as previous; renamed PEREZVON (see above)

## 7. Animals in Beatrix Potter

The children's picture books by the English writer Beatrix Potter (1866–1943) have been a firm favorite since their original appearance in the earliest years of the 20th century. The following list gives the type of animal unless it already forms part of the name. The names are listed in alphabetical order by first word or title (e.g., MRS.), even though many of the animals have "surnames."

AUNT PETTITOES, aunt of PIGLING BLAND
BENJAMIN BUNNY, a rabbit
DUCHESS, a little black dog
GINGER, a tomcat
GOODY TIPTOES, a gray squirrel, wife of TIMMY TIPTOES
HUNCA MUNCA and TOM THUMB, "two bad mice" (Chapter 12)
JEMIMA PUDDLEDUCK, a duck
JEREMY FISHER, a frog
JOHN JOINER, a dog
JOHNNY TOWN-MOUSE
LITTLE PIG ROBINSON
MR. ALDERMAN PTOLEMY TORTOISE
MR. JACKSON, a fat toad
MR. JOHN DORMOUSE
MR. TOD, a fox
MOPPET AND MITTENS, a pair of kittens, sisters of TOM KITTEN
MRS. CHIPPY HACKEE, a chipmunk
MRS. RIBSTON ("RIBBY"), a cat, cousin of MRS. TABITHA TWITCHIT

MRS. TABITHA TWITCHIT, a cat
MRS. THOMASINA TITTLEMOUSE, a mouse
MRS. TIGGY-WINKLE, a hedgehog
OLD MR. BOUNCER, a rabbit, father of BENJAMIN BUNNY
PETER RABBIT, with siblings FLOPSY, MOPSY, and COTTONTAIL
PICKLES, a terrier
PIGLING BLAND, a "sedate little pig," with brothers ALEXANDER, CHIN-CHIN, and STUMPY
PIG-WIG, "a perfectly lovely little black Berkshire pig"
SALLY HENNY-PENNY, a hen
SAMUEL WHISKERS, a rat, with wife ANNA MARIA
SIMPKIN, a cat
SIR ISAAC NEWTON, a newt
SQUIRREL NUTKIN
STUMPY, a large brown dog
TIMMY TIPTOES, a fat gray squirrel, husband of GOODY TIPTOES

| | |
|---|---|
| TIMMY WILLIE, a country mouse | TWITCHIT |
| TIPKINS, a small dog | TOMMY BROCK, a badger |
| TOM KITTEN, son of MRS. TABITHA | |

The names range from the ordinary words for the animals, including folk or nursery names, to human first names or surnames, and are often a combination of both kinds. Many of the characters were based on real pets that the author had when young. Among them were rabbits named PETER and BENJAMIN, a white rat SAMMY, and a hedgehog, MRS. TIGGY. MRS. TIGGY-WINKLE was thus based on the real hedgehog MRS. TIGGY, so called from *tiggy* as a folkname for this animal, in turn perhaps related to *tig* as a dialect word for a small pig. One book is devoted to the six FLOPSY BUNNIES. They are not individually named, but are the children of BENJAMIN BUNNY and his cousin FLOPSY, sister of PETER RABBIT ("I do not remember the separate names of their children," writes the author). MRS. RIBSTON seems to have a name derived from *Ribston pippin* as a type of apple. TIMMY WILLIE is so named as he is *timid*.

It should be noted that SQUIRREL NUTKIN is a British red squirrel, but TIMMY TIPTOES is an American gray one. A rabbit named PETER RABBIT occurs in other children's books, including *Old Mother West Wind* (1910) by the American writer Thornton W. Burgess, and the comic strip *Peter Rabbit* (1920) based on this character by the American illustrator Harrison Cady.

## 8. Animals in "Saki"

"Saki" was the penname of H.H. Munro (1870–1916), famous for his amusingly ironic stories of English society life in the years just before World War I (in which he was killed). His many short stories feature animals of all kinds, from pampered pets to racehorses, from farm animals to creatures that are half-animal, half-human. Not for nothing is one collection of 36 short stories entitled *Beasts and Superbeasts* (1914), with the stories themselves bearing titles such as "The She-Wolf," "The Boar-Pig," "The Hen," "The Stalled Ox," and "The Philanthropist and the Happy Cat." One of his best-known animals is the ferret who gave his name to the story "Sredni Vashtar," in the collection *The Chronicles of Clovis* (1911). The writer himself was a cat owner and lover, and tells of his admiration for the animal in "The Achievement of the Cat" in *The Square Egg* (1924).

The stories are full of witty comments and entertaining talk, which frequently touch on the name of an animal when it is first mentioned. Even so, there are many unnamed animals in Saki's writings, and it must be said that the animal names are in general much more conventional than those of the human characters, which are in many cases eccentric or surreal. (James Cushat-Prinkly, Alethia Debchance, Octavian Ruttle, and Crispina Umberleigh are typical examples.) Below is a listing of all named animals, together with relevant quotes concerning some of them.

ANNE DE JOYEUSE, a "handsome plum-roan gelding"

ATTAB, a "large tabby-marked cat" (from the region of Baghdad that gave the name of the tabby itself; see *tabby* in Appendix I, page 183)

BETTER NOT, a racehorse

THE BROGUE, "a light-weight hunter," otherwise "a useful brown gelding" The animal had been christened Berserker in the earlier stages of its career; it had been rechristened the Brogue later on, in recognition of the fact that, once acquired, it was extremely difficult to get rid of.

CLOVER FAIRY, "a young bull with a curly red coat"

COUNT PALATINE, a racehorse

"You think," said the Other, "that a name should economize description rather than stimulate imagination?"

"Properly chosen, it should do both. There is my lady kitten at home, for instance; I've called it Derry."

"Suggests nothing to my imagination but

DAISY, a cow

DERRY, a kitten

protracted sieges and religious animosities. Of course, I don't know your kitten —"

"Oh, you're silly. It's a sweet name, and it answers to it — when it wants to. Then, if there are any unseemly noises in the night, they can be explained succinctly: Derry and Toms."

(The references are to *Derry* or *Londonderry,* in Northern Ireland, scene of many sieges, and the former London department store, *Derry and Toms.*)

DON TARQUINIO, a Persian cat (compare TARQUIN SUPERBUS, below) His pedigree was as flawlessly Persian as the rug [...]. The page-boy, who had Renaissance tendencies, had christened him Don Tarquinio. Left to themselves, Egbert and Lady Anne would unfailingly have called him Fluff, but they were not obstinate.

ESMÉ, a hyena

"What on earth are we to do with the hyæna?," came the inevitable question.

"What does one generally do with hyænas?" I asked crossly.

"I've never had anything to do with one before," said Constance.

"Well, neither have I. If we even knew its sex we might give it a name. Perhaps we might call it Esmé. That would do in either case."

HARTLEPOOL HELEN, a game fowl, favorite of next

HARTLEPOOL'S WONDER, a gamecock

LE FIVE O'CLOCK, a French racehorse

LOUIS, a "diminutive brown Pomeranian"

LOUISA, "a rather fine specimen of the timber-wolf"

MOTORBOAT, a racehorse

"May I ask," said Mrs. Packlehide, amid the general silence, "why you put your

MYRTLE, a cow

NURSERY TEA, a racehorse

OAKHILL, a racehorse

OLD BIDDY, "a nice little roan cob"

OLD SHEP, a "white-nozzled, stiff-limbed collie"

PIPECLAY, a racehorse

POLLY, a parrot

SADOWA, a racehorse

money on this particular horse?" [...]

"Well," said Lady Susan, "you may laugh

at me, but it was the name that attracted me. You see, I was always mixed up with the Franco-German war; I was married on the day that the war was declared, and my eldest was born the day that peace was signed, so anything connected with the war has always interested me. And when I saw there was a horse running in the Derby called after one of the battles in the Franco-German war, I said I *must* put some money on it."

SNOW BUNTING, a racehorse
SREDNI VASHTAR, "a large polecat-ferret" (Chapter 12)
STARLING CHATTER, a racehorse
TARQUIN SUPERBUS, a "huge white Yorkshire boar-pig." (*Tarquinius Superbus* was a semilegendary 6th-century B.C. king of Rome.)
TOBERMORY, a talking cat
WHITEBAIT, a racehorse
WUMPLES, a Persian kitten

## 9. Animals in Wodehouse

The following animals, mainly dogs, occur in the novels and stories of the humorous writer P.G. Wodehouse (1881–1975). They are quoted from Geoffrey Jaggard, *Wooster's World* and Daniel Garrison's *Who's Who in Wodehouse* (see both in Bibliography). The type or breed of animal is stated.

ALEXANDER, a cat
ALPHONSE, a poodle
AMBROSE, a spaniel
AUGUSTUS, (1) a cat; (2) another cat
BARTHOLOMEW, an Aberdeen terrier
BILL, a fox terrier
BOTTLES, a mongrel
CAPTAIN KETTLES, a cat
CLARKESON, a Pekinese
CUTHBERT, a cat
CYRIL, a cocker spaniel
EGBERT, a swan
EMILY, a Welsh terrier
EMPRESS OF BLANDINGS, a Berkshire pig (Chapter 12)
EUSTACE, a monkey
JABBERWOCKY, a pug
JACK, a mongrel
JOSEPH, a black cat
LEONARD, a parrot
LYSANDER, a bulldog
McINTOSH, an Aberdeen terrier
MIKE, an Irish terrier
MITTENS, a sheepdog
THE MIXER, a terrier
NIGGER, a dog
PERCY, (1) a cat; (2) a dog; (3) a swan
PING-POO, a Pekinese
PIRBRIGHT, a Pekinese
POMONA, a Pekinese
POPPET, a dachshund
ROBERT, a spaniel
ROLLO, a bull terrier
SAM GOLDWYN, a mongrel
SIDNEY, a snake
SUSAN, a Pekinese
TIBBY, a cat
TOMMY, a pug
TOTO, a dog
WILHELM, an Alsatian (German shepherd)
WING-FU, a Pekinese

# APPENDIX IV : ZOO NAMES

The 152 names below were those of animals in the London Zoo, England, in the late 1920s, a hundred years after its founding by the London Zoological Society in 1827. Its initial collection of animals gradually grew in number and diversity, and in 1835 its first chimpanzee, TOMMY, arrived, prompting the following lines by the humorist Theodore Hook:

> The folks in town are nearly wild
> To go and see the monkey-child
> In gardens of Zoology
> Whose proper name is Chimpanzee.
> To keep this baby free from hurt
> He's dressed in a cap and Guernsey Shirt;
> They've got him a nurse and he sits on her knee
> And she calls him her Tommy Chimpanzee.

The most famous inmate of the London Zoo was undoubtedly the elephant JUMBO (see Chapter 10, page 120), acquired in 1865. The Zoo receives no state support, unlike many other famous zoos, such as the National Zoological Park, Washington, D.C., whose running costs are met by public funds, and its income is solely from admission charges, membership subscriptions, and gifts and bequests. (A steady decline in visitors, slumping from two million in 1968 to half this number in 1988, resulted in the near closure of the Zoo, the oldest in the world, in 1992.) The names are taken from Helen M. Sidebotham, *Round London's Zoo* (see Bibliography).

ABDULLA, a lion, brother of FATIMA
ALPHONSE, a genet (catlike animal related to civet)
ARTHUR, an ape
BABS, a cassowary
BABY, a hornbill

BABY KIANG, a baby kiang (Tibetan wild ass)
BARBARA, a polar bear
BEATTIE, a seal
BESSIE, (1) a kinkajou (honey bear); (2) a wild horse, mate of BOBBY, mother of SAMSON

## Appendix IV. Zoo Names

BETTY, (1) a bison, daughter (with JOAN) of PUNCH; (2) a reindeer, mate of JOE
BIDDY, a vixen
BILL, a dromedary
BILLY, (1) a goat; (2) a grizzly; (3) a kangaroo; (4) a llama, mate of MERICA; (5) a sealion; (6) a toucan; (7) a warthog
BLASTEW, a caracal (desert lynx), brother of DAMEUGH, both so facetiously named for aggressiveness
BLÜCHER, a rabbit; presumably for the Russian general (1889-1938), then in the news, rather than the Prussian field marshal (1742-1819)
BOB, a hippopotamus, mate of JOAN
BOBBY, a wild horse, mate of BESSIE, father of SAMSON
BOGEY, a black jaguar, so named from sinister appearance
BOOBOO, a (female) performing chimpanzee
CAROLINE, a lioness, mate of JUJA
CHANG, a Burmese elephant
CHARLIE, a zebra
CLEOPATRA, a python; presumably for the Egyptian queen who killed herself with an asp (a cobra, which is venomous, unlike a python, which is not)
COCKY, a cockatoo
CUPID, a camel
DAISY, (1) a buffalo, mate of PETER; (2) a clouded leopard; (3) a leopardess
DAMEUGH, a caracal (desert lynx), sister of BLASTEW (which see for origin of name)
DIANA, a hippopotamus; the animal bathed, like her mythological namesake
DIBS, a mangabey (white-eyelid monkey)
DIGGER, a kangaroo; from the nickname for an Australian
ELIZABETH, a rhinoceros
EUSTACE, an echidna
FATIMA, a lioness, sister of ABDULLA
FAY, a lioness, mate (with LENA) of TOTO
FELIX, a rhinoceros

FIREWORKS, a baby zebra; perhaps for combination of liveliness and dazzling black and white stripes
GEORGE, (1) an alligator; (2) a mandrill; name was used in zoo for all mandrills, perhaps for "manly" associations; (3) a pelican
GILGIL, a cheetah
GINGER, a camel
GLADYS, a hornbill
GYPSY, a brown bear
HANGO, a (female) African elephant; perhaps for a local placename
HARRY, a raven
HECTOR, a brown bear
JACK, (1) a brush turkey, mate of JILL; both so called as they went up and down a hill, like nursery rhyme characters; (2) a chimpanzee; (3) a lion club
JACKIE, (1) a Capuchin monkey; (2) a grizzly
JACKO, a Barbary ape
JENNIE, (1) a nilgai (Indian antelope); (2) a takin (related to the goat); (3) a mountain zebra, mate of JOE
JENNY, a rhesus monkey
JILL, a brush turkey, mate of JACK (which see for origin of name)
JIM, a puma
JIMMIE, (1) a performing chimpanzee; (2) a hippopotamus, son of BOB
JOAN, (1) a bison, daughter (with BETTY) of PUNCH; (2) a hippopotamus, mate of BOB
JOCK, a dingo
JOE, (1) a mountain zebra, mate of JENNIE; (2) a meerkat (South African mongoose); (3) a reindeer
JOEY, (1) a parrot; (2) a porcupine
JUBA, a cheetah
JUJA, a lion, brother of TOTO, mate of CAROLINE
KATHLEEN, a rhinoceros; "the rhino was nicknamed Kenya until she became the property of the Zoo, but then it was decided that a pet rhino ought to have a more friendly name, so she was renamed Kathleen" (page 193)
KIBERENGE, a (male) African

elephant; perhaps for *Kiberege (sic),* Tanzania
KITCHENER, a black bear, a former mascot of one of Lord *Kitchener's* regiments
KITTY, a marten
KLONDIKE, a wolf, mate of LOLLOPY
LADY, a reindeer, daughter of JOE
LAURA, a parrot
LENA, a lioness, mate (with FAY) of TOTO
LIZZIE, a polar bear, mate of SAM
LOLLOPY, a (female) wolf, who *lollops*
MAC, (1) a cassowary; (2) a golden eagle
MAGGIE, a giraffe
MARCUS, a giant ape
MARMADUKE, a tortoise
MAUD, a black leopard
MAUDIE, a giraffe
MAY, a puma, mother of JIM
MERICA, a (female) llama; presumably for animal's native South *America*
MICKY, (1) a (female) bison, mate of PUNCH; (2) a reindeer, son of JOE
MIKEY, a (female) wolf
MINNIE, a Burmese elephant
MOSES, a colobus (type of African monkey)
NANCY, a warthog
NELL, (1) a brown bear; fell in barrel of tar; so named for now forgotten character "Tar-Pot Nell"; (2) a puma
NELLIE, a giant ape, mother of MARCUS
NOSEY, an ant-eater; for the animal's prominent nose
OLD DAN, a vulture
PADDY, a giraffe
PEGGY, a small chimpanzee
PENNY, a penguin
PERCY, a hippopotamus, mate of DIANA
PETER, (1) a buffalo; (2) a small elephant; (3) an otter; (4) a woolly monkey
PETER NEIL, a tiger
PETER PAN, an alligator who "wouldn't grow up" and who shared a home with the crocodile WENDY, like the children in the J.M. Barrie play (1904) (in which the villainous pirate Captain Hook is eaten by a crocodile)

PIGGY, a pig-tailed monkey
POLA, a macaque
PONGO, an orangutan; from native African (Kongo) word for anthropoid ape
PUCK, a gibbon; name suggests a mischievous nature
PUNCH, a bison
RAJAH, a large tiger; named with title of Indian prince or king (compare RANEE)
RANEE, (1) an Indian elephant; (2) a tigress; both named with title of Indian princess or queen (compare RAJAH)
RANJI, a tigon (tiger-lion hybrid); so named as bred in India on estate of Rajput nobleman Prince *Ranjitsinhji* (1872–1933), today best remembered as an outstanding cricketer
RENIE, a prairie wolf
REX, a leopard, mate of DAISY
SALLY, a chimpanzee
SAM, (1) a black-maned lion, brother of SIMBA; (2) a polar bear, mate of LIZZIE; (3) another polar bear
SAMSON, a wild horse; presumably for biblical character (*Sampson*), famous for hair and strength
SATAN, a black leopard; so named from color
SHEBA, a cheetah
SIMBA, a black-maned lion, brother of SAM; from Swahili *simba,* "lion"
SOPHY, an ostrich
SUMBA, a Komodo dragon (kind of lizard), companion of SUMBAWA; for neighboring islands *Sumba* and *Sumbawa,* Lesser Sunda Islands, Indonesia
SUMBAWA, a Komodo dragon (kind of lizard), companion of SUMBA (which see)
SUSIE, a polar bear, mate of TROWSERS
TEDDY, a tiger cub; for its resemblance to *teddy* bear
TODDY, an otter; name suggests pet form of *otter* itself
TOMMY, a caracal (desert lynx)
TOTO, (1) a chimpanzee; (2) a lion cub, brother of JUJA

*Appendix IV. Zoo Names*

TROWSERS, a polar bear, mate of SUSIE; presumably for long-haired hind legs

TUBBY, a jaguar; so named from plumpness

WENDY, a crocodile; shared a home with PETER PAN (which see)

WHITE-TIP, a possum

WINNIE, a black bear; bear was origin of fictional WINNIE-THE-POOH (see Chapter 12, page 165)

# APPENDIX V : CELEBRITY PETS

The main section of this book has mentioned animals owned by heads of state and other famous people, many of them no longer alive. By contrast, below is a listing of animals, mainly pets, that are or have been owned by a selection of exactly 100 current or recent celebrities, almost all familiar from television. The person's profession is given together with their year of birth (where known), which may be relevant for the particular animal name.

Some celebrities are prolific animal owners. The American actress Kirstie Alley (born 1955), for example, has a veritable menagerie of 40 pets, including cats, dogs, fishes, birds, and a chicken. Alas, space is at a premium for them all to be listed here. Nor does it follow that the named animals below are the only ones owned by the particular person. The pop singer Michael Jackson is famous for his esoteric collection of pets, tantamount to a private zoo, but only a "sampler" is given below. The British actress Liza Goddard has five horses, five dogs, two cats, two chickens, and two bullocks. But only her favorite horse is named.

Where the breed of animal is known it is stated. Otherwise the basic animal is given. In a few cases the origin or particular reference of the name is supplied, where it is available.

Holly Aird (actress, 1969): dog, TOFFEE; cat, BOGGY
Mary Kay Ash (businesswoman, ?): poodle, GIGI
Eileen Atkins (actress, 1934): cats, ARCHIE, FINNEGAN, GUS
Amanda Bairstow (actress, 1960): dog, GEORGE
Roy Barraclough (actor, 1935): Highland White terrier, WHISKY

Allyce Beasley (actress, 1954): dog, RALPH
Shari Belafonte-Harper (actress, ?): dogs, RES, CAUSE
Gina Bellman (actress, 1966): cat, BOBO
Rodney Bewes (actor, 1938): cats, BERYL, PERCY
Erma Bombeck (columnist, 1927): Yorkshire terrier, MURRAY

Appendix V. Celebrity Pets 205

Tony Britton (actor, 1924): cats, TIGGER, HARRY
Judy Brooke (actress, 1970): crossbreed retriever/spaniel, BEN
Mel Brooks (actor, 1926): Staffordshire bull terrier, PONGO
Patrika Brown (businesswoman, ?): Lhasa apso, DANA, crossbreed, WOLFESS
Susan Brown (actress, 1946): cat, LUCY
Malandra Burrows (actress, 1965): mongrel, BONKERS
Michelle Byatt (actress, 1970): cat, BENJI
Simon Callow (actor, 1949): boxer, BRUNGE
Fanny Carby (actress, ?): cat, MRS. TIGGY
Anna Carteret (actress, 1942): black cat, MICHAEL JACKSON
Tracy Childs (actress, 1963): retriever, HARVEY
Julie Dawn Cole (actress, 1957): schnauzer, SHAMBLES
Nicola Cowper (actress, 1967): dog, SIDNEY
Olivia de Havilland (actress, 1916): Airedale, SHADRACH
Phyllis Diller (comedienne, 1917): Lhasa apso, PHEARLESS
Michael Kirk Douglas (actor, 1944): Jack Russell terrier, REGGIE
Jo Durie (tennis player, 1960): cat, PICKLES
Carmen du Sautoy (actress, 1952): cat, LUCKY
Rene Enriquez (actor, ?): dogs, GORDO (Spanish = "fat"), FLACO (Spanish = "thin")
Barry Fantoni (journalist, jazz musician, cartoonist, 1940): cocker spaniels, ARCHIE, JELLY ROLL MORTON
Marsha Fitzalan (actress, ?): mongrel, SOCKS
Eileen Fulton (actress, ?): shih tzu, SARAH BERNHARDT, Pekinese, SIR LAURENCE OLIVIER
Patricia Garwood (actress, 1941): cocker spaniel, CASSIE; cat, JOSIE
Jill Gascoine (actress, 1937): Labrador, ELLA; cats, DODGER, LIZ

William Gaunt (actor, 1937): cocker spaniel, RUPERT
Robyn Gibbes (actress, 1957): border collie, JAKE; cat, PUSS
Richard Gibson (actor, 1954): cats, BILL, NINE
Liza Goddard (actress, 1950): horse, OLIVE
Otto Graham (football player, 1921): mutt, MUFFIN, schnauzer, LITTLE OTTO
Veronica Hamel (actress, 1943): cat, BLACK
George Hamilton IV (actor, 1939): German shepherd, LOBA, beagle, DELILAH
Harry Hamlin (actor, 1951): cat, MENO
Marie Helvin (fashion model, 1952): cats, SHEBA LEE, SUSU
Bob Hope (actor, 1903): Alsatian shepherd, SNOW JOB, German shepherd, SHADOW
Susan Howard (actress, ?): dog, MOE
Terri Howard (actress, ?): cats, BOOBY, DAFFODIL, EMMA, KITTONE (later DINKY), LILY, RUSSELL; finches, UNCLE DEE DEE, DUM DUM (female)
Celia Imrie (actress, 1952): cat, MILDEW
Michael Jackson (pop singer, 1958): chimp, BUBBLES; giraffe, JABBAR; boa, MUSCLES; llama, LOUIE
Rachel James (actress, 1956): dog, MATTI BEAN
Louise Jameson (actress, 1951): dog, DIXIE; cat, SHEESH
Sue Jenkins (actress, ?): cats, WHISKEY, CHARLIE
Jasper Johns (artist, 1930): mongrel, WHISKEY
Diane Keaton (actress, 1946): cat, BUSTER KEATON
George Kennedy (actor, 1925): cairn terrier, BLYME; Maltese, BUTTONS
Jean-Claude Killy (skier, 1943): French shepherd, INDIAN
Eartha Kitt (singer, actress, 1928): poodles, MATTY, ABANAZA; cats, KIZZY, LUNNY, RAGS
Maxine Klingibaitis (actress, 1964): dog, HENRIETTA
Stepfanie Kramer (actress, 1956): dog, MAGGIE

Kim Lewis (actress, 1963): parrot, MR. POINTYBIRD
Shona Lindsay (actress, 1969): cat IVOR (for musician *Ivor* Novello)
Julian Lloyd Webber (cellist, 1951): turtles, BOOSEY and HAWKES (for music publishers), HODDLE and WADDLE (for two footballers)
Judy Loe (actress, 1947): dog, CASSIE; cats, RITA, ZIZ
Linda Lusardi (ex-model, presenter, 1960): dog, TRINA
Rue McClanahan (actress, 1935): dogs, SANDY, MISTY, HARROD; cats, FOSDICK CELESTINE
Cathy Rigby McCoy (gymnast, ?): cocker spaniel, MOLLY
Patrick Malahide (actor, 1945): springer spaniel, TOBY
Suzy Mallery (businesswoman, ?): terrier, HAPPY
Barry Manilow (singer, 1946): beagles, BAGEL, BISCUIT
Sally Ann Matthews (actress, 1970): dog, ALEX
Lisa Maxwell (singer, dancer, actress, ?): Tibetan spaniel, FIZZGIG
Paula Kent Meehan (businesswoman, ?): Irish setter, DUBLIN
Brian Moll (actor, 1925): Siamese cat, MITZIKA
David Morse (actor, ?): golden retriever, LUCY
Karen Murden (actress, 1970): cat, SYLVESTER
Martina Navratilova (tennis player, 1956): miniature fox terrier, K.D. ( = "killer dog")
Dame Anna Neagle (actress, 1904–1986): cat, TUPPENCE
Jack Nicklaus (golfer, 1940): golden retrievers, BEAR, LADY, WHITE PAWS
Mike Oldfield (pop musician, 1953): cat, SAUCEPOT
Susan Penhaligon (actress, 1949): dog, BRONWYN
Siân Phillips (actress, 1934): Burmese cats, BARNABY, RUPERT, SPENCER
Alan Plater (writer, 1935): red setter, THE DUKE (for Duke Ellington)
John Ratzenberger (actor, 1947): dogs, PETE, SALLY
Renee Richards (tennis player, 1934): Airedale, TENNIS-EE, Labrador, HOUDINI
Sharon Kay Ritchie (Miss America 1956, ?): Belgian shepherd, RANU
Pamela Salem (actress, ?): dog, JOSEPH
Prunella Scales (actress, 1932): cats, LILY, BAYLIS (i.e., LILY 'n' BAYLIS, for British theatre manager, founder of Old Vic, *Lillian Baylis*)
Sir Harry Secombe (singer, 1921): cat, MORIARTY
Dinah Sheridan (actress, 1920): shih tzu, COLEY
Sam Snead (golfer, 1912): Doberman pinscher, ADAM
Susanne Stuart (actress, ?): cat, PETUNIA
Gordon Thomson (actor, ?): dogs, JACK, LILY
Marilyn Van Derbur (Miss America 1958, ?): mongrel, BENJI
Andy Warhol (artist, 1928–1987): dogs, FAME, FORTUNE
Oliver Warman (artist, 1932): dogs, SALLY, ZETA, HERCULES, SEBASTIAN
Amanda Whitehead (actress, 1965): dog, HARRY
Geoffrey Whitehead (actor, 1939): dog, WALTER
Andy Williams (entertainer, 1930): boxer, BARNABY
Susannah York (actress, 1942): King Charles spaniel, ARCHY
Jacklyn L. Zeman (actress, ?): German shepherds, RUFFIAN, RUNNER
Efrem Zimbalist, Jr. (actor, 1923): German shorthair pointer, ZIMMIE

# APPENDIX VI : MOVIE TITLES INCLUDING ANIMAL NAMES

The names of various real or fictional animals have been promoted by their occurrence in the titles of feature movies. A selection of some of the better known such titles is given below, together with the film's country of origin, year of release, and identity of animal(s), where this is not obvious from the title itself. Additional information is given as appropriate, and (in cases where there is not a general cross reference to an earlier page) includes the publication year of a book on which a movie is based, especially where the book today is just as familiar as the resultant movie, possibly even more so. It can be assumed that the movie has the same title as the book unless a different original title is given.

In some instances the movie evolved from an existing cartoon character, either one in a comic strip, such as POGO or SNOOPY, or one in a short animated cartoon, such as PORKY PIG.

Animated cartoons have a boldface "c" after the release year.

*The Adventures of Gallant Bess* (US, 1984), rodeo horse
*Adventures of Rex and Rinty* (US, 1935), horse and dog
*Alakazam the Great* (Japan, 1960c), monkey
*Alf, Bill and Fred* (GB, 1964c), duck and dog (second and third names)
*An Alligator Named Daisy* (GB, 1955)
*Babar: The Movie* (Canada/France, 1989c), elephant (see page 142)
*Balthazar* (France/Sweden, 1966), donkey
*Bambi* (US, 1942c), forest deer (see page 142)
*Basil, The Great Mouse Detective* (US, 1986c) (GB title; US omits name)
*Bedtime for Bonzo* (US, 1951), chimpanzee
*Benji* (US, 1974), *Benji the Hunted* (US, 1987), and *For the Love of Benji* (US, 1977), mongrel dog
*Big Red* (US, 1962), Irish setter
*Bill and Coo* (US, 1947c), birds
*Black Beauty* (US, 1946; GB, 1971), horse (see page 143)
*Charlotte's Web* (US, 1973c), spider (E.B. White novel, 1952)
*Clarence the Cross-Eyed Lion* (US, 1965) (pilot for *Daktari* TV series)
*Crin Blanc* (France, 1953), wild horse (French = "white mane")
*Digby: The Biggest Dog in the World* (GB, 1973), old English sheepdog

*Dumbo* (US, 1941c), elephant (see page 148)
*Dusty* (Australia, 1982), half-wild dog (Frank Dalby Davison novel, 1946)
*Felix the Cat: The Movie* (US, 1989c) (see page 148)
*A Fish Called Wanda* (US, 1988) (see page 164)
*Flipper* (US, 1963), dolphin
*Fluffy* (US, 1964), lion
*Francis* (US, 1950), talking mule (all sequels include name)
*Fritz the Cat* (US, 1972c), New York alley cat
*Gallant Bess* (US, 1946), horse
*The Great Dan Patch* (US, 1949), racehorse
*Greyfriars Bobby* (GB, 1960) (see page 118)
*Gus* (US, 1976), mule
*Hambone and Hillie* (US, 1983), dogs
*Harry and Tonto* (US, 1974), cat (second name)
*Harvey* (US, 1950), rabbit (see page 150)
*Heathcliff: The Movie* (US, 1986c), cat
*Hey There, It's Yogi Bear* (US, 1964c) (see page 165)
*Hugo the Hippo* (US, 1976c)
*Ichabod and Mr. Toad* (US, 1949c) (second name only; see page 136)
*Jonathan Livingston Seagull* (US, 1973) (Richard Bach novel, 1970)
*Jumbo* (US, 1962), elephant (see page 120)
*K-9* (US, 1988), German shepherd dog
*Kelly and Me* (US, 1956), dog
*Kes* (GB, 1969), kestrel (Barry Hines novel *A Kestrel for a Knave*, 1968)
*King Kong* (US, 1933) and *Son of Kong* (US, 1933; sequel), giant ape (see page 153)
*Lady and the Tramp* (US, 1955c), pedigree spaniel and mongrel dog
*Lassie Come Home* (US, 1943), *The Magic of Lassie* (US, 1978; a remake), and *Son of Lassie* (US, 1945; sequel to *Lassie Come Home*), collie dog (see page 153)
*The Legend of Lobo* (US, 1962), forest wolf (Spanish = "wolf")

*Matilda* (US, 1978), kangaroo (Paul Gallico novel, 1970)
*Mickey's Christmas Carol* (US, 1983c), mouse (see page 154)
*Mighty Joe Young* (US, 1949), giant gorilla (follow-up to *King Kong*, above)
*Moby Dick* (GB, 1956), white whale (see page 164)
*My Friend Flicka* (US, 1943) and *Thunderhead, Son of Flicka* (US, 1945; sequel to *My Friend Flicka*), horse (see page 155)
*My Pal Trigger* (US, 1946), horse (see page 131)
*Napoleon and Samantha* (US, 1972), lion (first name)
*Nikki, Wild Dog of the North* (US, 1961), wolf dog
*Old Yeller* (US, 1957), dog (see page 155)
*Oliver and Company* (US, 1988c), kitten (Dickens novel *Oliver Twist*, 1837)
*Orca — Killer Whale* (US, 1977)
*Owd Bob* (GB, 1938), sheepdog (see page 155)
*Perri* (US, 1957), squirrel
*Phar Lap* (Australia, 1983), racehorse (see page 126)
*Pogo for President* (US, 1984c), possum
*Porky Pig in Hollywood* (US, 1986c)
*Rhubarb* (US, 1951), ginger cat
*Savage Sam* (US, 1962), dog (in original novel, son of *Old Yeller*, above)
*Scudda Hoo, Scudda Hay* (US, 1948), mules
*Silver Blaze* (GB, 1937), racehorse (see page 160)
*Smoky* (US, 1946), cow horse (Will James novel, 1926)
*Snoopy Comes Home* (US, 1973c), beagle (see page 161)
*The Story of Seabiscuit* (US, 1949), racehorse (GB title: *Pride of Kentucky*)
*Tonka* (US, 1958), horse
*Turner and Hooch* (US, 1989), dog (second name)
*The Voice of Bugle Ann* (US, 1936), hound (Mackinlay Kantor novel, 1935) (see page 144)

*Appendix VI. Movie Titles Including Animal Names*

*Who Framed Roger Rabbit* (US, 1988, part c) (see page 137)
*Won Ton Ton, the Dog Who Saved Hollywood* (US, 1976), German shepherd

# BIBLIOGRAPHY

The following is a selection of titles, mainly books but also articles, devoted wholly or partly to animals and their names, sometimes factually, sometimes anecdotally. A few of the titles concern themselves directly with the naming process, even if only by way of recommendation or suggestion. Other titles are more literary in character, such as Michael Joseph's *Charles: The Story of a Friendship*. Even so, they are worth reading for the names they contain, and for the stories behind them.

Aberconway, Christabel (comp.). *A Dictionary of Cat Lovers*. London: Michael Joseph, 1949.
Alexander, David. *A Sound of Horses*. Indianapolis: Bobbs-Merrill, 1966.
Amaral, Anthony. *Movie Horses*. Indianapolis: Bobbs-Merrill, 1967.
Anderson, Janice. *The Cat-a-logue*. Enfield: Guinness Books, 1987.
*The Animals Who's Who*. London: Marshall Cavendish, 1973.
Annett, Bill, and Marta Annett (comps.). *For the Love of Animals: True Stories from the Famous*. London: Arrow Books, 1989.
Baker, Stephen. *5001 Names for Cats*. New York: McGraw-Hill, 1983.
Bartlett, Kay. "It May Be a Dog's Life, but Pooch's Name Is Its Own," *Detroit News*, June 19, 1977.
Bentley, Nicolas, Michael Slater, and Nina Burgis. *The Dickens Index*. Oxford: Oxford University Press, 1988.
Brewer, E. Cobham. *The Dictionary of Phrase and Fable*. Leicester: Blitz Editions, 1990 [1894].
Browder, Sue. *The Pet Name Book*. New York: Workman Publishing, 1979.
Browne, Michèle. *The Royal Animals*. London: W.H. Allen, 1981.
Bryant, Mark. *The Complete Lexicat: A Cat Name Compendium*. London: Robson Books, 1992.
Bryant, Traphes. *Dog Days at the White House*. New York: Macmillan, 1975.
Caras, Roger A. *A Celebration of Cats*. London: Robson Books, 1989.
———. *The Roger Caras Dog Book*. San Diego: Holt, Rinehart & Winston, 1980.
———. *Treasury of Great Cat Stories*, rev. ed. London: Robson Books, 1990.
———. *Treasury of Great Dog Stories*. London: Robson Books, 1990.
Carmichael, Pamela. *The SHE Book of Cats*. London: Ebury Press, 1983.
Carpenter, Humphrey, and Mari Prichard. *The Oxford Companion to Children's Literature*. Oxford: Oxford University Press, 1984.
Clark, Cumberland. *The Dogs in Dickens*. London: Chiswick Press, 1926.

Collison, Robert L. *A Jorrocks Handbook*. London: Coole Book Service, 1964.
Connor, William. *Cassandra's Cats*. London: Hutchinson, 1958.
Cook, E. Thornton. *Sir Walter's Dogs*. Edinburgh: Grant & Murray, 1931.
Cooper, Jilly. *Animals in War*. London: Heinemann, 1983.
Croft-Cook, Rupert, and Peter Cotes. *Circus: A World History*. New York: Macmillan, 1976.
Crowell, Pers. *Cavalcade of American Horses*. New York: McGraw-Hill, 1951.
Daly, Macdonald. *Royal Dogs*. London: W.H. Allen [circa 1953].
Davis, Jim. *The Garfield Book of Cat Names*. New York: Ballantine Books, 1988.
Denenberg, R.V., and Eric Seidman. *Dog Catalog*. New York: Grosset & Dunlap, 1978.
Detrick, Mary Helen, and Nancy Butler White. *Naming Your Pet*. New York: Arco Publishing, 1979.
Downey, Fairfax. *Great Dog Stories of All Time*. Garden City, N.Y.: Doubleday, 1962.
D'Oyley, Elizabeth. *An Anthology for Animal Lovers*. London: William Collins & Son, 1927.
Durant, John, and Alice Durant. *Pictorial History of the American Circus*. Cranbury, N.J.: A.S. Barnes, 1957.
Evans, Ivor H. *Brewer's Dictionary of Phrase and Fable*. London: Cassell, 14th ed., 1989.
Farson, Daniel. *In Praise of Dogs*. London: Harrap, 1976.
Fenin, George N., and William K. Everson. *The Western*. New York: Crown Publishers, 1962.
Fireman, Judy. *Cat Catalog*. New York: Workman Publishing, 1976.
Fogle, Bruce. *Pets and Their People*. London: Collins, 1983.
Freeman, William. *Everyman's Dictionary of Fictional Characters*, 3rd ed., revised by Fred Urquhart. London: J.M. Dent, 1973.
Garrison, Daniel H. *Daniel Garrison's Who's Who in Wodehouse*. New York: Peter Lang, 1987.
Gettings, Fred. *The Secret Lore of the Cat*. London: Grafton Books, 1989.
Gibson, Frank A. "Dogs in Dickens," *The Dickensian*, September 1957.
Gilbert, John R. *Cats*. London: Paul Hamlyn, 1961.
Goodman, Jonathan. *Who He?* London: Buchan & Enright, 1984.
Greaves, John. *Who's Who in Dickens*. London: Hamish Hamilton, 1972.
Haddon, Celia. *Faithful to the End: A "Daily Telegraph" Anthology of Dogs*. London: Headline, 1991.
Hall, Angus. *Monsters and Mythic Beasts*. Garden City, N.Y.: Doubleday, 1976.
Halliwell, Leslie. *Halliwell's Film Guide*, 8th ed., edited by John Walker. London: HarperCollins, 1991.
———. *Halliwell's Filmgoer's Companion*, 9th ed. London: Grafton Books, 1988.
Hamilton, Elizabeth. *In Celebration of Cats*. Newton Abbot: David & Charles, 1977.
Hardwick, Michael, and Mollie Hardwick. *The Charles Dickens Encyclopedia*. Reading: Osprey, 1973.
Hare, C.E. *The Language of Sport*. London: Country Life, 1939.
Harvey, Sir Paul (comp. and ed.). *The Oxford Companion to English Literature*, 4th ed. Oxford: Clarendon Press, 1967.
Hodder-Williams, Ernest (ed.). *My Dog Friends*, "by the author of 'Where's Master?'" London: Hodder & Stoughton [1915].
Husband, M.F.A. *A Dictionary of the Characters in the Waverley Novels of Sir Walter Scott*. London: George Routledge & Sons, 1910.
Jaggard, Geoffrey. *Wooster's World*. London: Macdonald, 1967.
Johnes, Carolyn Boyce. *Please Don't Call Me Fido*. New York: Berkley Publishing, 1977.
Johns, Rowland. *Dogs You'd Like to Meet*. London: Methuen, 1924.

Joseph, Michael. *Cat's Company*. London: Michael Joseph, 1946.
———. *Charles: The Story of a Friendship*. London: Michael Joseph, 1943.
Kandel, Thelma. *What to Name the Cat*. New York: Linden Press/Simon & Schuster, 1983.
Kelly, Niall. *Presidential Pets*. New York: Abbeville Press, 1992.
Kirk, Mildred. *The Everlasting Cat*. Woodstock, N.Y.: Overlook Press, 1977.
Kolatch, Alfred J. *Names for Pets*. Middle Village, N.Y.: Jonathan David Publishers, 1971.
Lenburg, Jeff. *The Encyclopedia of Animated Cartoons*. New York: Facts on File, 1991.
Lewis, Martyn. *Cats in the News*. London: Macdonald Illustrated, 1991.
———. *Dogs in the News*. London: Little Brown, 1992.
Longrigg, Roger. *The History of Horse Racing*. New York: Stein & Day, 1972.
Loxton, Howard (ed.). *Dogs*. London: Paul Hamlyn, 1962.
MacGregor, Alastair. *Cat Calls*. London: Weidenfeld & Nicolson, 1988.
Mason, James, and Pamela Mason. *The Cats in Our Lives*. London: Michael Joseph, 1949.
Maxwell, Bede C. *The Truth About Sporting Dogs*. London: Pelham Books, 1972.
Mayle, Peter, and Arthur Robins. *Anything but Rover!* London: Arthur Barker, 1985.
Mercatante, Anthony S. *Zoo of the Gods*. New York: Harper & Row, 1974.
Moore, Daphne. *The Book of the Foxhound*, 2nd ed. London: J.A. Allen, 1974.
Nichols, Beverley. *Beverley Nichols' Cats ABC*. London: Jonathan Cape, 1960.
———. *Beverley Nichols' Cats XYZ*. London: Jonathan Cape, 1961.
Palmer, Joan. *Animals: All Famous, Mostly Friendly*. Melksham: Colin Venton, 1974.
Parker, Eric, and A. Croxton Smith. *The Dog-Lover's Week-End Book*. London: Seeley Service [1950].
Passage, Charles E. *Character Names in Dostoyevsky's Fiction*. Ann Arbor: Ardis, 1982.
Philip, Alex. J. *A Dickens Dictionary*. London: George Routledge & Sons, 1909.
Porter, Valerie. *The Guinness Book of Almost Everything You Didn't Need to Know About Dogs*. Enfield: Guinness Books, 1986.
———. *Horse Tails*. Enfield: Guinness Books, 1989.
Potter, Stephen, and Laurens Sargent. *Pedigree: Words from Nature*. London: Collins, 1973.
Pring, Peter. *Winners of the World's Major Races from 1900*. Sydney: Thoroughbred Press, 1982.
Pringle, David. *Imaginary People*. London: Grafton Books, 1987.
Pullein-Thompson, Josephine. *Horses and Their Owners*. London: Nelson, 1970.
*Registered Names of Horses*. Wellingborough: Weatherbys, 1984.
Reid, Beryl. *The Cat's Whiskers*. London: Ebury Press, 1986.
Ritvo, Harriet. "Pride and Pedigree: The Evolution of the Victorian Dog Fancy," *Victorian Studies* (Indiana University), vol. 29, no. 2, winter 1986.
Room, Adrian. *Pet Names*. Bognor Regis: New Horizon, 1979.
Ross, Estelle. *The Book of Noble Dogs*. London: Century Co., 1922.
Rothel, David. *The Great Show Business Animals*. San Diego and New York: A.S. Barnes, 1980.
Schinto, Jeanne (ed.). *Of Dogs and Men*. London: André Deutsch, 1991.
Seymour-Smith, Martin. *The Dent Dictionary of Fictional Characters*. London: J.M. Dent, 1991.
Sheard, Wilfred. *The Glory of the Dog: Some Tales from the Story of the Dog in History and Legend*. London: Hutchinson, n.d. (c. 1935).
Shook, Carrie, and Robert L. Shook. *What to Name Your Dog*. New York: Tribeca Communications, 1983.
Sidebotham, Helen M. *Round London's Zoo*. London: Herbert Jenkins, 1928.

Simpson, Frances. *The Book of the Cat.* London: Cassell, 1903.
Smith, Bradley. *The Horse in the West.* Cleveland: World, 1969.
Smith, Donald J. *Horses at Work.* Wellingborough: Patrick Stephens, 1985.
Stern, Jane, and Michael Stern. *Amazing America.* New York: Random House, 1977.
Stuart, D.M. *A Book of Cats: Legendary, Literary and Historical.* London: Methuen, 1959.
Summerhays, R.S. (comp.). *Summerhays' Encyclopedia for Horsemen.* London: Frederick Warne, 1952.
Taggart, Jean E. *Pet Names.* Metuchen, N.J.: Scarecrow Press, 1962.
Thurber, James. "How to Name a Dog," in *The Beast in Me, and Other Animals.* London: Hamish Hamilton, 1949.
Tibballs, Geoff. *The Golden Age of Children's Television.* London: Titan Books, 1991.
Tremain, Ruth. *The Animals' Who's Who.* London: Routledge & Kegan Paul, 1982.
Truman, Margaret. *White House Pets.* New York: David McKay, 1969.
Vevers, Gwynne (comp.). *London's Zoo.* London: Bodley Head, 1976.
Walker, Stella A. *Horses of Renown.* London: Country Life, 1954.
Watson, J.N.P. (comp.). *The World's Greatest Dog Stories.* London: Century Publishing, 1985.
Weekley, Ernest. *An Etymological Dictionary of Modern English.* New York: Dover Publications, 1967 [1921].
_____. *Words and Names.* London: John Murray, 1932.
Wels, Byron G. *Animal Heroes.* New York: Macmillan, 1979.
Whitfield, June. *Dogs' Tales.* London: Robson Books, 1987.
Winter, Gordon. *The Horseman's Week-End Book.* London: Seeley Service, n.d. [1937].
Wood, Gerald L. *The Guinness Book of Animal Facts and Feats,* rev. ed. Enfield: Guinness Superlatives, 1976.

# INDEX

*This index contains most of the names in the main section of the book, some 2,700 of them, but not those in any of the appendices. It also does not include the names of animals serving to create the name(s) of their offspring, typically as the sire and dam of a racehorse. The index specifies type of animal without regard to age ("cat" versus kitten"), except in the case of farm animals, where distinctions between young and adult animals ("cow" or "calf," "sheep" or "lamb") have commercial value. Gender is treated similarly: a "dog" could be a bitch, but cows and bulls, as farm animals, are differentiated. As the book also considers the generic names of some animals that derive from a personal name, these are also indexed, but in italics.*

A.J. *cow* 101
A.P. Indy *horse* 105
Abaster *horse* 134
Abatos *horse* 134
Abba *starling* 32
Aberdeen Lassie *horse* 84
Able *monkey* 105
Able Lunda *calf* 103
Above Suspicion *horse* 85
Abracadabra *dog* 74
Abraxas *horse* 134
Absent-Minded Beggar *cat* 79
Abu Al-Abbas *elephant* 105
Achilles *dog* 94; *horse* 93
Actaeon *horse* 133
Adam's Pet *horse* 84
Admiral *horse* 83
Adonis *dog* 69
Aethon *horse* 133
Aeton *horse* 134
Age of Consent *horse* 85
Agnes *calf* 99
Agrippina *cat* 105
Ah Choo *cat* 79
Ah Choo Lucky *dog* 37
Air Power *horse* 82
Airborne *horse* 87
Airy-Fairy *cat* 28
Ajax *horse* 93; *rabbit* 60
Akela *wolf* 141

Al Borak *horse* 134
Albert the Alligator *alligator* 137
Alexander *cat* 30, 60
Alexander the Mouse *mouse* 138
Alexandra *dog* 58
Alf's Carino *horse* 84
Algy Pug *dog* 138
Alice *lamb* 101
Alidoro *dog* 141
Alikat *cat* 38
Alma North *horse* 86
Alpha *calf* 101
Alsvid *horse* 134
Alverstone Tiggy *dog* 70
Always Faithful *horse* 83
Always Happy *horse* 83
Am, Ch, Salisyn's Macduff *dog* 72
Amber *cow* 98; *dog* 24; *general* 23; *pony* 8
Amethea *horse* 133
Amoroso *dog* 68
Amos *cat* 51
Amy *dog* 73
An Mee Too *dog* 23
Anchor *horse* 93
Ancient *cat* 58
Andante *dog* 68

Andie *cat* 58
Andy *cat* 54
Andy Panda *panda* 137
Angel *goldfish* 59
Angus *dog* 30, 58, 60
Animal *cat* 64
Anna *dog* 30
Annie *cow* 100
Annie Rooney *dog* 74
Anorexia *pig* 101
Anthony *dog* 58; *pig* 19
Apis *bull* 134
Apollinaris *cat* 106
Apollyon *horse* 141
Apple of My Eye *horse* 84
Apricot *cow* 98
April *cow* 99
Arabia *cow* 97
Arc de Triomphe *dog* 68
Archer *dog* 88
Arctic Slave *horse* 85
Arctophonos *dog* 133
Ardent *dog* 88
Ardrose *horse* 88
Arduous *dog* 88
Argentan *horse* 83
Argus *dog* 88, 133
Ariel *cow* 101; *dog* 88
Arielstar *dog* 68
Aries *cat* 28

215

# Index

Arion *horse* 134
Arkle *dog* 89; *horse* 106
Armlet *dog* 88
Aroni *dog* 57
Arsenal *dog* 88
Arsenic *dog* 88
Artaxerxes *horse* 93
Artful *pig* 100
Arthur *goldfish* 42
Article *dog* 88
Artless *dog* 88
Arundel *horse* 141
Arvak *horse* 134
Ashbocker *cat* 42
Ashe *dog* 32
Aslan *lion* 134
Asta *dog* 142
Aster *cow* 98, 110
Astronaut *dog* 112
Athalass *horse* 88
Atlas *pony* 25
Atossa *cat* 106
Attila *dog* 94
Aubretia *cat* 26
Audrey *cow* 100
Aul' Mither *sheep* 101
Aunt Norah *cow* 99
Aure-U-Lupi *horse* 85
Australia *cow* 97
Autocrat *dog* 74
Auto-Speed *horse* 82
Averof *horse* 85
Awful Affliction *dog* 74

B.M. *horse* 102
B Major *horse* 81
B.T. *cat* 27
Ba-Ba *dog* 28
Baba *cat* 79
Babar *elephant* 142
Babbacombe Bumble *dog* 70
Babe *dog* 74
Babieca *horse* 142
Baboushka *horse* 32
Baby *cat* 62
Baby Buggins *dog* 74
Babylon *dog* 61
Badger *badger* 136; *cat* 25
Bagheera *panther* 142
Bagpuss *cat* 138
Baiardo *horse* 142
Bailer *dog* 68
Baker *monkey* 105
Balaclava Boy *horse* 85
Baldies *hens* 102
Balios *horse* 134
Baloo *bear* 142
Baltic *cow* 98
Bambi *deer* 142
Bandit *dog* 6
Bandoola *elephant* 106
Bang *dog* 70

Bar Gold *horse* 83
Bar Silvero *horse* 83
Barbara *chimpanzee* 49
Barbican *horse* 93
Barbmark *horse* 88
Barmy *horse* 86
Barney *dog* 6
Barney Bear *bear* 137
Barre-de-Rouille *cat* 106
Barrow Boy *horse* 84
Barry *dog* 106
Basha *dog* 31
Basil *cat* 61
Basil Brush *fox* 137
Basket *dog* 107
Bass *cat* 59
Battleship *dog* 92
Baviaan *baboon* 142
Bavieca *horse* 142
Baxter *dog* 63
Baxter the Wall *dog* 63
Bay *cat* 61
Bayard *horse* 142
Be Friendly *horse* 83
Be True *horse* 83
Beaky *budgerigar* 28
Bean *horse* 43
Bear *dog* 6
Beau *dog* 107
Beau Sovereign *horse* 82
Beautiful Joe *dog* 107
Beauty *horse* 40, 100; *pony* 8
Beauty Boy *cat* 79; *dog* 131
Bébé *cat* 115
Beelzebub *cat* 106
Beerbohn *cat* 107
Beethoven *cat* 34
Behave Too *horse* 83
Behi's Csions Csiny *dog* 72
Beine *dog* 28
Belaud *cat* 107
Belka *dog* 107
Bell *hen* 101
Bellevue *cat* 27
Bellyn *ram* 14, 15
Belter *hamster* 31
Belzebub *horse* 142
Ben *dog* 6, 7, 8, 9, 46, 84;
  *goldfish* 8, 9; *rat* 8, 9
Ben Botherit *dog* 74
Bendix *cat* 40
Benedicta *cat* 43
Bengy *dog* 7
Benji *dog* 6, 8, 107
Beppo *dog* 63
Bergamot *cat* 61
Berganza *dog* 142
Bernard *cat* 58
Bertha *cow* 99; *dog* 29
Bertie *cat* 27
Beryl *cow* 99
Beryl the Peril *pig* 101
Betsy *hen* 101

Betty *pig* 100
Bevis *dog* 143
Bianco *horse* 24
Big Ben *dog* 121
Big Bertha *cow* 99
Big Boy *dog* 115
Big Philou *horse* 87
Bigwig *rabbit* 143
Bijou *dog* 15
Bilbo *dog* 42
Bildad *dog* 5
Bilge *dog* 68
Bill *lamb* 100; *mouse* 9
Bill Badger *badger* 138
Bill XXII *goat* 107
Billy *budgerigar* 8, 9; *cat* 54,
  124; *dog* 49, 62; *goat* 13,
  18; *stick insect* 9
Billy Boy *budgerigar* 31
Billy Girl *budgerigar* 31
Bingo *dog* 143
Bingo the Brainy Pup 138
Binkie *dog* 143
Birdsnest Sally *cat* 78
Birdsnest Sam *cat* 78
Biscow *cat* 52
Biscuit *pony* 9
Biscuit Boy *horse* 84
Biscuits *horse* 34, 37
Bitch *fish* 32
Bitser *dog* 29
Bitter *cat* 59
Bizzie Lizzie *cat* 34
Black Agnes *horse* 107
Black Beauty *horse* 136, 143
Black Cygnet *horse* 83
Black Regent *horse* 82
Black Saladin *horse* 108
Blackberry *dog* 125; *rabbit* 151
Blackbird *horse* 100
Blackie *cat* 7, 23; *dog* 6, 126;
  *rabbit* 8; *rat* 8
Blacky *cat* 23
Blair *dog* 21
Blanch *dog* 144
Blanche *cat* 155; *pig* 24
Blanco *dog* 24, 108
Blatherskite *cat* 106
Blemie *dog* 108
Bless This Horse *horse* 83
Blitz *dog* 94; *horse* 82
Blitzen *general* 59
Block *dog* 68
Blondi *dog* 108
Blossom *cat* 53
Blow *dog* 74
Blue Acre *horse* 83
Blue Emperor *cat* 48
Bluebell *cow* 99
Blueboy *cat* 76
Blueman *dog* 91, 92
Bluey *budgerigar* 8, 9; *cat*
  43; *cow* 98; *general* 23

## Index

Blushing Maid *horse* 85
Blyth *cat* 61
Boatswain *dog* 108
Bob *dog* 70, 94; *gerbil* 60
Bobbity Bumpkins *dog* 31
Bobby *budgerigar* 8, 9; *calf* 99; *rabbit* 46
Bodane *horse* 88
Bodger *dog* 108
Bogus *dog* 52
Bold Bidder *horse* 86
Bold Experience *horse* 86
Bold Hour *horse* 86
Bold Lad *horse* 86
Bold Love *horse* 85
Bold Reasoning *horse* 86
Bold Ruler *horse* 86
Boldini *horse* 85
Boldnesian *horse* 86
Bolero *dog* 68; *horse* 39
Bombalurina *cat* 136
Bonk *cat* 34
Bonnie *cat* 61; *dog* 6, 8
Bonnington *cat* 34
Bonny *calf* 103; *dog* 92; *horse* 100
Bont *dog* 57
Bonzo *chimpanzee* 22, 108; *dog* 22, 51, 138
Boo *cat* 63; *lamb* 101
Boo-Boo *bear* 144
Booboo *chimpanzee* 144
Boom-de-ay *dog* 74
Bootie *cat* 26
Bootiful *cat* 58
Boots *dog* 144
Bootsie *sheep* 103
Booty *cat* 26
Boozer *dog* 91
Borage *cat* 61
Borak *horse* 134
Boredom *pig* 101
Boric *dog* 62
Boss Cat *cat* 138, 163
Boston Beans *dog* 125
Bounce *dog* 108; *horse* 100
Bouncer *dog* 73
Bouquet *cow* 98
Boxer *cat* 41; *dog* 144; *horse* 100, 144
Boxy *dog* 2
Boy *dog* 108
Boycott *general* 10
Bracelet *dog* 89
Bracken *dog* 10, 89; *pony* 24
Brackett *cat* 59
Bramble *cow* 99; *horse* 51; *pony* 8
Bran *dog* 135
Brandy *dog* 6, 47, 57, 94
Brandy Snap *cow* 98
Brave Lad *horse* 83
Brec *horse* 42

Brer Bear *bear* 144
Brer Fox *fox* 144
Brer Rabbit *rabbit* 144
Brer Tarrypin *turtle* 144
Brer Wolf *wolf* 144
Brian *snail* 147
Brigade Major *horse* 83
Brigadier Gerard *horse* 109
Brigadore *horse* 135, 144
Brigand *dog* 89
Brigham *horse* 53, 109
Brigliadoro *horse* 144
Brillo *dog* 68
Broadway Dancer *horse* 86
Broadway Girl *horse* 84
Broken Tail *cat* 27
Bronte *horse* 133
Brookwood *cat* 43
Broom *dog* 68
Brown Jack *horse* 109
Brownie *cow* 99; *dog* 6, 23, 160
Browny *general* 23
Bruce *budgerigar* 58; *cat* 46; *dog* 94
Bruin *bear* 14
Bruiser *dog* 84
Bruno *dog* 10
Brush *dog* 63, 68
Bryn *pony* 42
Bryny *cat* 28
Bryony *cat* 61
Bubble *gerbil* 8, 9; *guinea pig* 8, 9, 47; *lamb* 103; *mouse* 8, 9
Bubbles *gerbil* 8, 9; *goldfish* 8, 9; *guinea pig* 8, 9; *hamster* 8, 9; *pony* 41
Buccaneer *horse* 31
Bucephalus *horse* 1, 109
Buck *dog* 144, 165
Bucket *dog* 31, 68
Buckthorn *rabbit* 151
Buddy *dog* 6, 109
Buffy *dog* 6
Bugle Ann *dog* 144
Bugle Boy *horse* 84
Bugs Bunny *rabbit* 137
Bugsy *rabbit* 8, 9
Bulger *cat* 79
Bulldozer *cat* 31
Bullet *dog* 109
Bull's-Eye *dog* 145
Bullwinkle *moose* 138
Bumble *cat* 32; *guinea pig* 23
Bumpkin *cat* 64
Bunbury *cat* 43
Bungey *cat* 64
Bunny *cow* 97
*bunny rabbit* 22
Burmese *horse* 110

Bursar *dog* 73
Bury *hatchet fish* 30
Bushy *dog* 63
Buster *lamb* 100
Bustifer Jones *cat* 64
Bustopher Jones *cat* 64
Butch *dog* 6, 31
Buttercup *cow* 99; *dog* 74; *pig* 100
Butterfly *dog* 126
Buttie *goat* 35
Buttons *dog* 6
Byerly Turk *horse* 110


Cable *dog* 88
Cactus *cow* 98
Cadbury *cat* 24
Cadger *dog* 74
Cadpig *dog* 145
Caesar *dog* 110, 122
Calamity Jane *dog* 125
Calf-Chaser *cow* 99
Calvin *cat* 110
Cameo *cat* 79
Camera *dog* 88
Camouflage *horse* 86
Camp Commander *horse* 83
Canberra *dog* 68
Candour *dog* 88
Candy *cow* 99; *dog* 6, 23
Capilet *horse* 145
Capsule *dog* 88
Captain *dog* 131; *horse* 100, 101
Captain Flint *parrot* 145
Captain Pugwash *cat* 49
Cara *cat* 58
Card King *horse* 82
Caro *cat* 42
Carol *dog* 37, 63
Carole *cow* 99
Carpenter *dog* 33
Carraway *dog* 34
Carrie Ann *dog* 39
Carroty *general* 23
Carruthers *cat* 45
Caruso *canary* 110; *cat* 10
Casamayor *horse* 94
Casanova *swordfish* 32
Casher *cat* 24
Caspar *horse* 39
Caspian *dog* 88
Cassius *cat* 29; *dog* 39
Cassy's Pet *horse* 84
Castor *cat* 59
Cat *cat* 145
Catamanda Cavalcade *cat* 78
Catarina *cat* 110
Catkin *cat* 27
Catty *cat* 39; *catfish* 30

Cavalier *cat* 59; *dog* 57
Cavall *dog* 145
Ceasar (*sic*) *dog* 94
Cecil *lamb* 101
Celer *horse* 111
Cerberus *dog* 133
Ceres *dog* 68
Cerus *horse* 134
Chamberlain *dog* 92
Chamois *dog* 68
Chamomile *cat* 61
Champ *dog* 6
Champion *horse* 111
Chamy *horse* 24
Chanoine *cat* 115
Chanticleer *cock* 14, 15, 101
Charcoal *cow* 98
Charles *cat* 111; *dog* 72; *lamb* 101
Charles O'Malley *cat* 111
Charley *cat* 46; *dog* 31, 111; *pigeon* 45; *pony* 111
Charley-Girl *dog* 31
Charlie *budgerigar* 8, 9; *bull* 99; *cat* 7; *dog* 6, 7, 10, 94; *goldfish* 46; *horse* 39; *pig* 101; *stick insect* 9
Charlie Boy *dog* 45
Charlotte *spider* 145
Charmer *horse* 43
Charybdis *donkey* 59
Chaucer *cat* 27
Checkers *dog* 111
Chee Chee *monkey* 145
Cheeky *dog* 74
Cheri *dog* 8
Cherokee Lad *pony* 26
Cherry *cat* 58; *general* 23; *pony* 25
Chervil *cat* 61
Chesara Chervil of Sedora *dog* 72
Cheshire *cat* 33
Chestnut *horse* 23
Chew *cat* 58
Chia-Chia *panda* 112
Chiang *dog* 41
Chico *dog* 6
Chicory *cat* 61
Chiffy *cat* 58
Chil *kite* 146
Childebrand *cat* 115
Chin Chin *dog* 74
Ching-Ching *panda* 112
Chinkapen *dog* 112
Chinnie *cat* 79
Chip an' Dale *chipmunks* 138, 146
Chipmunk *dog* 68
Chippendale *dog* 35
Chloe *dog* 7
Chocolate *calf* 99
Cholmondeley *cat* 31; *chimp* 112
Chopin *cat* 35
Christian *dog* 30
Christmas *cat* 52
Christopher *cat* 52
Chua *dog* 43
Chuffy *cat* 38, 58
Chum *dog* 74
Chumly *cat* 79
Chump *dog* 74
Chunkie *dog* 5
Churchill *dog* 92
Cicely Audrey Coffeebean Kaulback *goat* 112
Ciel Rouge *horse* 83
Cigarette *dog* 74
Cincinnati *horse* 112
Cinderella *dog* 38, 57
Cinders *cat* 25
Cindy *cat* 25; *dog* 6; *general* 23
Circumstance *horse* 93
Circus *cow* 98
Clanger *foal* 27
Claude *cat* 45
Claude Cat *cat* 137
Clavileno *horse* 146
Cleat *dog* 68
Clementine *horse* 41
Clever Pal *horse* 83
Climax *cat* 79
Clipper *dog* 68, 126
Clive *cat* 54
Clockwork *pony* 31
Cloudy *cat* 40
Clova *dog* 42
Clover *cow* 99; *horse* 146
Clown *pony* 25
Cluneen Merry Minuet *dog* 71
Cluneen Merry Minuet of Littlecourt *dog* 71
Clyde *calf* 103
Coart *hare* 14, 15
Cobber *dog* 62
Cobweb *dog* 26
Coco *cat* 38; *dog* 5, 6, 112
Code of Love *horse* 85
Coffeebean *goat* 112
Cola *dog* 24
Colo *gorilla* 113
Colocynth *dog* 91
Colonel *dog* 146; *horse* 100
Colonel Chubby *dog* 74
Colonel Nelson *horse* 83
Colossus *dog* 61
Comanche *horse* 113
Comet *dog* 68
Comfortably Off *horse* 83
Comfrey *cat* 61
Commotion *dog* 74
Constellation *dog* 68
Convair *dog* 68
Convoy *dog* 92
Cookie *cat* 24
Cop *dog* 42
Copenhagen *horse* 113
Copper *dog* 92; *pony* 8, 23
Coral *horse* 47, 102
Cordon Rouge *horse* 83
Coriander *cat* 61
Coricopat *cat* 136
Cornelia *dog* 5
Cornflake *horse* 28
Cornwater Kirsty *cat* 75
Corry *dog* 39
Corsair *dog* 68
Cortina *cat* 54
Cos *cat* 61
Cottontail *rabbit* 61
Coup de Feu *horse* 82
Court Sensation *horse* 82
Courtoys *dog* 14
Crab *dog* 146
Cracker *cat* 40
Crackers *dog* 63
Cracksman *horse* 87
Craig *cat* 42
Cress *lamb* 103
Crevette *cat* 115
Crick *dog* 52
Cricketer *dog* 89
Crispin *horse* 39
Crown *horse* 93
Crown Case *horse* 82
Crumpet *cock* 101
Crusoe *dog* 136, 146
Cuddles *animal* 138
Cupid *dog* 68; *horse* 85
Cuppy *cat* 2
Curiosity *cat* 79
Curlews *cat* 43
Curley *cow* 97
Curlicue *dog* 27
Curtal *horse* 146
Cut *horse* 147
Cuwaert *hare* 15
Cwm *dog* 42
Cyder *dog* 57
Cylgal *horse* 88
Cyllaros *horse* 134

D-Day *dog* 92
D-Tail *cat* 79
Daffodil *cow* 99; *dog* 90
Daffy Duck *duck* 137
Dainty *cow* 98; *dog* 89
Dairymaid *dog* 89
Daisy *calf* 99; *cow* 99; *dog* 7, 10; *pig* 100
Daisy Duck *duck* 147
Dalal *elephant* 113

# Index

Damosel *dog* 92
Dancers Countess *horse* 86
Dancing Girl *horse* 84
Dangerous *dog* 90
Daph *dog* 146
Dapple *horse* 23
Darby *horse* 100
Darkie *cow* 97; *horse* 100
Darky *general* 23
Darley *horse* 122
Darley Arabian *horse* 113
Darling *horse* 100
Dash *dog* 113, 114
Dastardly *mouse* 33
Daub *duck* 60
David *dog* 41; *lamb* 103
Dawn Chorus *pigs* 102
Dawn Run *horse* 114
Dear Arthur *horse* 84
Debbie *cow* 99
Deborah *pig* 100
Decent *dog* 91
Dee and Me *horse* 84
Delight *goldfish* 59
Demon Path *horse* 82
Demus *cat* 58
Dennis *cat* 32
Dennis the Dachshund *dog* 137
Deputy Dawg *dog* 137
Derringham *cat* 43
Derry *cat* 43, 58
Derry D. *cat* 43
Desert Chat *horse* 85
Desert Orchid *horse* 114
Devil's Jump *horse* 42
Devonshire Cream *cat* 79
Dewdrop *dog* 90
Dewi *dog* 73
Di *dog* 147
Diamond *dog* 63, 147; *horse* 100
Diamond Lil *cat* 28
Diana *cat* 34; *dog* 61
Diana of Wildwood *dog* 125
Dice *pony* 41
Dick *cat* 31; *horse* 100
Dicky *cat* 26
*dicky bird* 13, 19
Dido *dog* 60
Didyme *cat* 64
Digger *dog* 42, 62
Dilly *guinea pig* 35
Dimity *cat* 79
Dimly *cat* 55
Dimple *cat* 47; *dog* 90
Dinah *cow* 99; *dog* 90
Dinba *cat* 54
Dinky Duck *duck* 137
Dinos *horse* 134
Dinty Pippin *pony* 52
Diogenes *dog* 147
Diomed *horse* 115

Dippy Dawg *dog* 138, 149
Disciplinarian *horse* 78
Dividend Bond *pig* 103
Dlinenky *cat* 128
Dobbin *horse* 13, 17, 147
Dr. Zhivago *horse* 39
Doggy *dog* 30
Dollar *cat* 42
Dollette *horse* 88
Dolly *calf* 103; *cow* 98, 99
Domino *horse* 24
Don Pierrot de Navarre *cat* 115
Donald Duck *duck* 137, 147
Donatello *tortoise* 9, 10
Donner *general* 59
Dookie *dog* 62, 63
Dora *cow* 99; *dog* 52
Doris *cow* 99
Dot *cat* 79
Dottie *pony* 25
Double Cheque *horse* 83
Doublet *calf* 103
Dougal *dog* 7, 147
Dovecote *dog* 90
Dragon *horse* 100
Dreamer *dog* 90
Drink-a-Pint *cow* 101
Drop-Even *horse* 86
Drummer *horse* 100
Duchess *dog* 6
Duff *dog* 72
Duffer *cow* 99
Duffy *dog* 25
Duke *dog* 6, 63, 115; *horse* 100
Dukie *dog* 63
Dumbo *elephant* 148
Dumpling *horse* 100
Dungheap *cow* 103
Dustbin *dog* 34
Duster *dog* 68
Dusty *dog* 6
Dylan *pony* 52; *rabbit* 147

Eagle *dog* 68
Eartha *dog* 48
Ebony *general* 23
Echo Esquire *dog* 68
Eclipse *horse* 115
Eddie *duck* 103
Edith *cat* 49
Edom *cat* 32
Edward *chimpanzee* 49
Edward Trunk *elephant* 138
Edwin *shark fish* 52
Eeyore *donkey* 148
Elaine *dog* 68
El Cid *goat* 107
El Morzillo *horse* 124
Electricity *dog* 74

Elegant *dog* 68
Elemauzer *cat* 135
Elfin *dog* 68
Elke *dog* 52
Ella *cat* 54
Elsa *dog* 8; *lion* 107
Elsie *cat* 54
Elvira *cat* 46
Emblem *dog* 108
Emerald *cow* 98
Emi *cat* 43
Emily *cat* 46
Emma *cat* 54; *cow* 99; *dog* 6, 7
Empress of Blandings *pig* 148
Emu *emu* 138
Ena *dog* 48
Enchantress *dog* 68
Enjolras *cat* 115
Enoch *gosling* 60
Eoos *horse* 134
Epi *horse* 35
Epistle *horse* 35
Eponine *cat* 115
Epsilon Bay *horse* 35
Eric *lamb* 103
Erica *dog* 73
Erl King *dog* 147
Ermintrude *cow* 12
Ernie *horse* 39; *lamb* 103
Eros *dog* 68
Erythreos *horse* 133
Esau *lamb* 103
Estralita *dog* 68
Ethel *cow* 99
Evening Star *dog* 68
Exemplary *horse* 83
Exploding *horse* 82

Fable *dog* 63
Fadda *mule* 134
Fair Cousin *horse* 84
Fair Tactics *horse* 83
Fairest *dog* 89
Fairy Queen *dog* 69
Fala *dog* 115
Famous *dog* 89
Fancy *dog* 89
Fanny *cat* 46
Fashoda *cat* 79, 80
Fate *cat* 40
Favourite *dog* 92
Fawn *cat* 79
Fay *cow* 99
Feather *dog* 89
Felix *cat* 148
Felix the Cat *cat* 137
Fell Swoop *horse* 85
Feller *dog* 116
Ferdinand *bull* 99, 148; *hamster* 60

# Index

Fergie *cat* 58
Fido *dog* 21, 47, 116, 132
Fiercy *gerbil* 60
Fiery Kiss *horse* 85
Fifi *dog* 8
Fileuse *horse* 87
Fillmore *dog* 49
Firefly *pony* 38
Firenza *cat* 54
First Grey *horse* 83
Fiver *guinea pig* 30; *rabbit* 148
Flagman *dog* 89, 90
Flame *dog* 24, 94
Flannel *dog* 68
Flapper *budgerigar* 34
Flash *dog* 94
Flasher *dog* 50
Flax *lamb* 100
Fleur *cat* 50; *dog* 50
Fleur de Lys *dog* 68
Fleur d'Or *horse* 83
Flick *pony* 26
Flicka *horse* 148
Flip the Frog *frog* 137
Flipper *goldfish* 8, 9
Floating Penny *horse* 83
Flopsy *rabbit* 8, 9, 61
Flora *cow* 99
Floreat Etona *dog* 74
Floss *horse* 34
Flossy *dog* 116
Flower *horse* 100
Flowerpot *budgerigar* 33
Fluffy *cat* 7, 8, 76; *dog* 6, 63; *guinea pig* 8; *hamster* 8; *rabbit* 8
Flush *dog* 116
Fly *cow* 97
Flying Childers *horse* 116
Flying Hero *horse* 82
Foinavon *horse* 116
Folies Bergeres *dog* 68
Folly *pony* 41
Follyfoot *horse* 37
Footy *cat* 54
Forest Fern *dog* 69
Forester *dog* 131
Fortune *cow* 100
Foss *cat* 116
Fourpence *cat* 39
Foxglove *cow* 99
Foxy *dog* 63, 125
Fozzie Bear *bear* 138
Francis the Mule *mule* 138
Franco *cat* 53; *dog* 92
Freaky Fanny *cow* 99
Freckles *dog* 127
Fred *goldfish* 8, 9, 46; *rabbit* 46; *stick insect* 8, 9; *tortoise* 8, 9, 10, 46
Fred Basset *dog* 138
Fred Tosh *goldfish* 28

Freddie *stick insect* 9
Freefoot *horse* 82
Freeman *cat* 61
Frenchie *dog* 25
Freya *cat* 59
Fricka *cat* 59
Friday *cat* 39
Friendly *gerbil* 60
Fritz *dog* 8
Frolic *horse* 100; *pony* 31
Frost *dog* 92
Fudge *cow* 98; *dog* 47
Funny *cat* 38
Funny Face *cow* 99
Funny Lad *horse* 84
Furette *horse* 88
Furry *cat* 54
Furry Wee *cat* 54
Fury *cat* 54; *dog* 94; *horse* 116

Galathe *horse* 134
Gale *dog* 31
Game Rights *horse* 86
Gamin *cat* 64
Gandhi *dog* 116
Gandy Goose *goose* 137
Gannet *cow* 97
Garfield *cat* 138, 148
Gargantua *gorilla* 116
Garm *dog* 149
Garryowen *dog* 149
Gato *horse* 117
Gavroche *cat* 115
Gay *cat* 61; *cow* 99
Gay Filou *horse* 87
Gaydier *horse* 88
Gazelle *dog* 69
Gazza *general* 10
Geist *dog* 117
Gelert *dog* 149
Gemini *dog* 29, 59
Gemma *dog* 6, 7
Gen *cat* 26
General *cat* 26
General Custer *horse* 83
Gentle Castra of Reeves *dog* 71
Gentle Castra of Reeves of Rathcondel *dog* 71
Gentle Thoughts *horse* 85
Gentleman in Black *dog* 74
Genuine *horse* 83
Geoff's Choice *horse* 84
George *budgerigar* 8, 9; *cat* 46; *chimpanzee* 49; *dog* 46; *giraffe* 117; *goldfish* 8, 9; *mouse* 9; *stick insect* 8; *tortoise* 9
Georgette *lamb* 100
Georgie Girl *cat* 55

Geraldine *cow* 99
Gerbella *gerbil* 30
Gertrude *kangaroo* 149
Ghost *dog* 74
Gib *cat* 17
Gideon *cat* 149; *dog* 50
Gifted *dog* 88
Giggle *dog* 88
Gilbert *dog* 88; *horse* 100
Gilbert O'Sullivan *guinea pig* 49
Gilda *dog* 63
Gilder *dog* 88
Gillian *dog* 69, 88
Gilly *cat* 42
Gimbal *dog* 88
Gimcrack *dog* 88, 89; *horse* 117
Gimlet *dog* 88
Gimmick *dog* 88
Gina Gay Kim *dog* 69
Ginger *cat* 7, 8, 57; *dog* 6, 88; *guinea pig* 8, 23
Ginky *guinea pig* 30
Ginny *cat* 26
Gip *cat* 117
Gipsy *dog* 88, 91
Girder *dog* 88, 91
Girdle *dog* 88, 91
Girlish *dog* 88
Gizmo *hamster* 9
Gladys *cow* 99
Glasgow *dog* 88
Gleaming *dog* 88
Gleeful *dog* 88
Glen *dog* 121
Glisten *dog* 88
Gnasher *dog* 160
Go Baby Go *horse* 82
Go Bang *dog* 73
Go for Broke *horse* 83
Go Too *horse* 82
Gobang *dog* 74
Gobbles *fish* 34
Godfrey *horse* 40
Godolphin Arabian *horse* 117
Godolphin Barb *horse* 117
Godzilla *dinosaur* 149
Goebals *dog* 64
Gold Coast *horse* 83
Gold Dust *horse* 43
Gold Tipped *horse* 83
Golden Beauty *horse* 83
Golden Miller *horse* 117
Golden Pippin *pony* 24
Goldie *eagle* 117; *goldfish* 8, 23
Goldsmith Maid *horse* 118
Goldylocks *cat* 79
Golfsticks *cat* 79
Goma *gorilla* 118
Good *cat* 61
Good Courage *horse* 83

## Index

Good Lad *dog* 73
Goofy *cat* 64; *dog* 138, 149
Gordon *cat* 31
Gospill Hill *horse* 85
Grace *cat* 155
Grace Darling *dog* 74, 75
Grani *horse* 134
Granny *sheep* 100
Great Love *horse* 85
Green Signal *horse* 83
Greenwood *dog* 90
Greg *dog* 30
Gregor *dog* 62
Gretchen V. Greif *dog* 72
Grey Shoes *horse* 83
Greyfriars Bobby *dog* 118
Greyhound *horse* 18
Griezell Greedigutt *cat* 135
Grimalkin *cat* 21
Grip *raven* 149
Grits *dog* 118
Grizabella *cat* 150
Grizzle *horse* 150
Grotesque *dog* 74
Grumps *dog* 146
Grymbert *badger* 14
Gub Gub *pig* 145
Gubbins *cat* 64
Guggle-Wumpf *cat* 64
Guinness *dog* 25
Guinsea *guinea pig* 65
Gulliver *pony* 31
Gunner *dog* 94
Gurkasses *ferret* 103
Gus *cat* 48, 150
Gustave *cat* 64
Gutter *blackbird* 33
Guy *gorilla* 110
Guzzie *dog* 34
Gybe *dog* 68


H.R.H. *dog* 74
Hachiko *dog* 118
Haggis *pony* 29; *tortoise* 29
Haig *cat* 26
Hairy King *dog* 74, 75
Hal *dog* 69
Half-Pint *dog* 25
Halle *horse* 39
Ham *chimpanzee* 118
Hambletonian *horse* 118
Hamish *cat* 55
Hamlet *hamster* 47; *pig* 100
Hamlyn *cat* 53
Hammy *hamster* 8, 9, 30
Handover *dog* 92
Handsome Major *horse* 83
Hank *budgerigar* 58
Hannah *cat* 52
Hardy *cat* 61; *horse* 83
Harlequin *dog* 69

Harmony *dog* 69
Harold *chimpanzee* 49
Harpagus *horse* 134
Harriet *gerbil* 35
Harry *dog* 7; *hamster* 8, 9; *mouse* 9
Harvey *dog* 151; *rabbit* 150
Hashish *cat* 34
Hathi *elephant* 150
Hazel *cow* 99; *general* 23; *rabbit* 150
Hazelnut *sheep* 100
He-She *cat* 31
Heartbeat *horse* 85
Heather *cow* 99; *dog* 63
Heavenly Pleasure *dog* 74
Heavenly Prince *horse* 82
Hecate *cat* 39
Heckle *turtle* 59
Heckle and Jeckle *magpies* 138
Hector *dog* 68
Heidi *dog* 6, 8, 119
Helleborium *dog* 92
Help *dog* 29
Hendrik *cat* 51
Henry *cat* 51, 62; *dog* 68; *hamster* 9; *sheep* 100
Hep *dog* 29
Hephzibah *dog* 29
Heps *dog* 29
Her *dog* 119
Herald *dog* 69
Herbert *pigeon* 45
Hercule *dog* 64
Hercules *cat* 59; *dog* 94; *hamster* 29
Herm *cat* 43
Hernia *dog* 91
Herod *cock* 57; *horse* 119
Heroine *dog* 92
Heron *cow* 97
Hervey *dog* 151
Hessian *lamb* 100
Hester *cow* 99
High *horse* 93
High and Interesting *dog* 73
High Award *horse* 83
High Wire *horse* 86
Hilda *cow* 99
Him *dog* 119
Hinge *cat* 59
Hinse of Hinsfield *cat* 119
Hippo *dog* 28
Hippocampus *horse* 134
Hippogriff *dog* 91
His Nibs *dog* 73
Hissing Syd *cat* 35
Hitler *dog* 64
Hodge *cat* 119
Holly *cat* 58; *dog* 10; *lamb* 103; *pony* 9; *whale* 10
Holst *tortoise* 40

Holt *cat* 135
Homer *hamster* 31
Homera *hamster* 31
Homespun *cat* 27
Honest John *fox* 151
Honesty *pony* 41
Honey *cow* 99; *hamster* 8, 9; *horse* 32
Honey Lover *horse* 85
Honey Suckle *dog* 73
Hooligan *dog* 90
Hoover *dog* 34, 63
Hope of Holland *horse* 85
Hoppy *cow* 99
Horrible *dog* 90
Horror *dog* 90
Horsey *horse* 30
Houdini *general* 33
Hovel *dog* 90
Hovis *cat* 26
How Nice *dog* 74
Howard the Duck *duck* 138
Howitzer *dog* 91
Hrimfaxi *horse* 134
Huckleberry Hound *dog* 137
Hudson Hammy *hamster* 45
Huey, Dewey and Louie *ducklings* 151
Hugh *cat* 47
Hugo the Hippo *hippopotamus* 137
Humbug *cow* 99
Humphrey *cat* 64, 119; *dog* 49
Humphrey Archibald Marmaduke Daniel "Fifi" Entancelin Shpeedy Gonzalez James Lewis Marshall Ellis *cat* 64
Hunca Munca *mouse* 151
Hurlyburlybuss *cat* 119
Hurry Round *horse* 82
Hush Money *horse* 83
Hyacinthus *horse* 86
Hycilla *horse* 86
Hyena *cow* 101
Hylander *horse* 86
Hypericum *horse* 86
Hyperides *horse* 86
Hyperion *horse* 86, 119
Hysteria *lovebird* 60


I *cat* 79
I, Beauty's Daughter *cat* 79
I Say *horse* 86
Iblis *cat* 115
Igloo *dog* 119
Ilemauzar *cat* 135
Imp *dog* 69
Imperial Crown *horse* 82
Incitatus *horse* 120

# Index

Infragummies Inokenunnies Infradilga Inclaynana Mickey Mouse Brittain *mouse* 65
Ingo *dog* 5
Ingraban *dog* 5
Ingulf *dog* 5
Inkerman *dog* 74, 75
Intrepid *dog* 131
Invisible Lad *horse* 84
Invisible Romance *horse* 85
Iris *cow* 99
Ironside *dog* 94
Iroquois *horse* 120
Irresistible Miss *horse* 84
Irrtum *horse* 85
Irving *dog* 69
Isaac *cat* 53
Isabella Bellamissa Schnopsa Brittain *mouse* 65
Isengrym *wolf* 13, 14
Italy *cow* 97
Its-a-Match *horse* 85
Itsy *cat* 29
Ivanhoe *dog* 69
Ivo *dog* 5
Ivory Lady *horse* 83
Ivy *cat* 58; *lamb* 103; *whale* 10
Izaak *cat* 34


J.C. *dog* 28
J. Fred Muggs *chimpanzee* 120
J. Worthington Foulfellow *fox* 151
Jabberwocky *cow* 98
Jacaranda *cow* 98
Jacey *dog* 28
Jack *lamb* 103
Jack Frost *cat* 79
*jackdaw* 13, 16, 18
Jackie *cow* 98
Jacko *monkey* 13, 16
Jack's Hope *horse* 84
Jacob *lamb* 103
Jacoba *cow* 98
Jacqueline *cow* 98
Jade *cat* 62; *cow* 98
Jael *cow* 98
Jam *cat* 24
James *cat* 48; *duck* 103
James Bond *pig* 103
James of Burgercroft *dog* 70
Jamie *dog* 7
Jane *cow* 98, 99
Janet *cow* 98, 99
Janice *cow* 98
Janine *cow* 98
Janstar *pony* 39
Japhet *cat* 47
Jaquetta *cow* 98

Jarmara *dog* 135
Jasmine *cow* 98; *duck* 103
Jason *dog* 6, 7, 43, 94
Jasper *cat* 62, 163
Jaws *goldfish* 8, 9
Jay *cow* 98
Jeckle *turtle* 59
Jeff Davis *horse* 131
Jefferson *dog* 49
Jehenna *cow* 98
Jellylorum *cat* 136
Jemima *cow* 98; *duck* 103
Jemima Puddleduck *dog* 34
Jemma *cow* 98
Jenny *cat* 26, 52; *cow* 99; *donkey* 13, 18; *duck* 46; *horse* 100; *mouse* 8
Jeoffrey *cat* 151
Jerry *general* 60; *gerbil* 8; *goldfish* 8; *lamb* 103
Jes *dog* 53
Jesca *cow* 98
Jess *dog* 152
Jesse *cow* 98
Jessica *cow* 98
Jessop *cow* 98
Jester *dog* 34; *lamb* 26
Jet *cow* 98; *dog* 70, 94; *general* 23; *horse* 100
Jewel *cow* 98; *dog* 92
Jib *dog* 68
Jig *cow* 97
Jill *cow* 99; *lamb* 103
Jill McSwine *dog* 73
Jillian *cow* 98
Jim *chimpanzee* 49; *sheep* 100
Jim Moo *cow* 101
Jiminy Cricket *cricket* 152
Jimmy *cat* 31
Jindy Lynn *dog* 69
Jingles *dog* 39
Jinjee *cat* 24
Jip *dog* 152
Jo *cat* 49
Jo-Jo *cat* 64
Joan *cow* 98
Joanna *cow* 98
Jocelyn *cow* 98
Jock *cat* 120; *dog* 152
Jocko *monkey* 16
Jocunda *cow* 98
Joey *budgerigar* 8, 9, 13, 17, 31; *canary* 13, 17; *gerbil* 9
John Keats *dog* 33
Jola *cow* 98
Jolly *dog* 63; *horse* 100
Jolly Jet *horse* 86
Jolly Lucky *horse* 83
Jonah *cat* 38
Jonathan *lamb* 103
Jonathan Livingston Seagull *seagull* 44, 152

Jonquil *cow* 98
Josepha *cow* 98
Josephine *cat* 62; *hamster* 61
Josie *budgerigar* 31; *cow* 99
Joy *cow* 98; *dog* 32
Juanita *cow* 98
Jub *dog* 62
Jubilee *cat* 79
Jubilee Girl *horse* 84
Judah *cow* 98
Judy *cat* 46; *cow* 99; *lamb* 103; *pig* 100
Juggernaut *dog* 94
Julia *cat* 61; *cow* 99
Juliana *cat* 61
Julie *cow* 98
Juliet *cat* 32, 61; *cow* 98
Juman *cow* 98
Jumble *dog* 152
Jumbo *dog* 37; *elephant* 120
Jumbo Jet *horse* 82
Junia *dog* 3
Juniper *cow* 98
Juno *cow* 98; *dog* 3, 92, 147
Jupiter *dog* 61
Just a Chance *horse* 83
Justin Morgan *horse* 121
Justina *cow* 98
Juvanescence *horse* 85


Kaa *python* 152
Kahn *dog* 94
Kaiser *dog* 73
Kali *calf* 99
Kalu *canary* 40
Kandida *dog* 69
Kanga *kangaroo* 136
Kanthaka *horse* 121
Karl *dog* 8
Karma *calf* 99
Kashtanka *dog* 152
Kate *cow* 99; *dog* 7
Kath *cat* 55
*katydid* 19
Keel *dog* 68
Keeper *dog* 121
Keith *cat* 59
Kelly *dog* 94
Kelpie *dog* 63
Kenneth *horse* 94
Kermit the Frog *frog* 138
Kernie *cat* 26
Kerrygold *dog* 69
Kestrel *cow* 98
Khan *dog* 74
Kiki-la-Doucette *cat* 152
Kilts *dog* 74
Kim *cat* 50; *cow* 99; *dog* 7, 94, 95
Kincsem *horse* 121
King *dog* 6, 92, 94, 95

## Index

King Arthur *horse* 84
King Boris *horse* 84
King Cole *dog* 126
King Cuthbert *horse* 84
King Frost *dog* 92
King Herod *horse* 119
King Kong *gorilla* 153
King Louis *lamb* 48
King Mausolus *dog* 61
King Max *horse* 84
King Midas *horse* 83
King of the Silver *cat* 79
King Samuel *horse* 84
King Silver *horse* 83
King Timahoe *dog* 121
King Tut *dog* 121
King's Flight *horse* 82
Kinky *dog* 27
Kirsty *calf* 99
Kiss *dog* 109
Kissing *horse* 85
Kit *horse* 100
*kittiwake* 19
Kitty *cat* 13, 16, 18, 30; *cow* 98
Kiwi *cat* 42
Knighterrant *dog* 69
Knirps *dog* 73
Knobbie *dog* 5
Kohinoor *dog* 74
Koko *gorilla* 121
Kola *cat* 79
Korky the Cat *cat* 137
Kosmos *dog* 74
Krazy Kat *cat* 137
Krishna *calf* 99
Kron Prinz *dog* 73
Kudryavka *dog* 121

La Mia Ragazza *horse* 84
Lacy *cat* 28
Laddie *dog* 94
Lady *dog* 6, 8, 92, 153; *horse* 28; *pony* 8, 32
Lady Goldsmith *horse* 118
Lady Golightly *dog* 74
Lady Lola *cat* 79
Lady Lollipop *cat* 79
Lady Pink *cat* 79
Ladybird *cat* 62
Ladyblush *dog* 73
Ladye Gaye *dog* 69
Laelaps *dog* 133
Laika *dog* 121
Lamb Chop *lamb* 138
Lampos *horse* 133
Lantern *pig* 100
Lares *dog* 58
Larry the Lamb *lamb* 137
Lassie *dog* 47, 72, 153; *pig* 100

Late Love *horse* 85
Laudable Pus *cat* 47
Laura *cow* 99; *pig* 100
Lauries Dancer *horse* 86
Lavinia *pig* 100
Lawman *dog* 69
Lawrence *dog* 50
Le Coq d'Or *horse* 83
Le Duc *horse* 82
Le Filou *horse* 86
Le Roi *dog* 74
Lea *horse* 43
Leah *pig* 100
Lennie *general* 2
Lenny the Lion *lion* 137
Leo *cat* 79; *lion* 13, 17
Leonardo *tortoise* 8, 9, 10
Lexington *horse* 122
Libella Labaja *cat* 76
Lido *cat* 40
Lie *cat* 58
Lien-Ho *panda* 122
Light Fingers *horse* 87
Light-O-Day *pig* 100
Lightning *general* 60; *goldfish* 31; *horse* 28, 93
Lilac *pig* 100
Lilac Wine *horse* 83
Lilith *cat* 115
Lily *cow* 100
Lilywood *dog* 90
Limonchik *dog* 121
Limsy *horse* 88
Linbury Lass *horse* 84
Lincoln *dog* 30
Linda *pig* 100
Lindy *horse* 111
Linger Longer Loo *dog* 73, 74
Linger Longer Lucy *dog* 74
Lingpopo *cat* 79
Linnet *cow* 98
Lion *dog* 153
Lippy the Lion *lion* 137
Lisa *dog* 8; *general* 2
Little Bo Bleep *horse* 85
Little Diamond *cow* 99
Little Man *cat* 62
Little Mouse *dog* 37
Littlehampton *cat* 43
Littlejohn *dog* 69
Littlewood *dog* 4
Lively *horse* 100
Lizetta *horse* 43
Lizzie *cat* 62; *cow* 99
Lolita *pig* 100
Lone Ranger *dog* 69
Lonesome *horse* 46
Lootie *dog* 122
Lord Cremorne *cat* 79
Lord Gwynne *cat* 79
Lorna *pig* 100
Lorraine *dog* 69

Lotus *cat* 54
Louisa *pig* 100
Lovely *dog* 153
Lovely Match *horse* 85
Lover Boy *cat* 34
Lovey *dog* 153
Luath *dog* 1, 122, 135
Lucifer *cat* 127
Lucinda *dog* 69
Lucky *cat* 8, 28, 40; *cow* 101; *dog* 6, 94, 153
Lucky Charm *dog* 69
Lucy *cat* 7, 8, 9, 28, 43; *dog* 6, 7, 8, 9; *cow* 99; *lamb* 100; *pig* 100
Ludovic le Cruel *cat* 127
Lufra *dog* 153
Luke *cat* 41; *dog* 7
Lulu *cat* 62; *cow* 99; *dog* 63, 74
Lunar Queen *horse* 82
Lundy *cow* 97
Lustre *pig* 100
Lydia *pig* 100
Lyman *dog* 91
Lyndale King *dog* 69

Mabel *cow* 99; *dog* 89
Mac *dog* 57
Macaroni *dog* 74; *pony* 122
McGregor *dog* 30
McGuirk *cat* 26
Madam *cow* 98; *dog* 91
Madame-Théophile *cat* 115
Madly Gay *horse* 83
Madness *dog* 91
Maera *dog* 133
Mafillette *horse* 84
Magdalena *dog* 39
Magic *budgerigar* 8; *cat* 33; *cow* 101
Magnanimous *horse* 83
Magnificat *cat* 29
*magpie* 13, 18
Magpie *cow* 98; *dog* 89
Mahatma Coatma Collar Gandhi *dog* 116
Mahogany *horse* 25
Maida *dog* 58, 122
Main *horse* 93
Maisie *cow* 99; *sheep* 103
Majesty *horse* 82
Major *dog* 94; *pig* 153
Major Role *horse* 83
Makarova *horse* 85
Malik *horse* 29
Malvina *seagull* 52
Malvolio *mynah bird* 26
Mamselle de Paris *dog* 68
Man Friday *lamb* 103
Man Kind *horse* 85

## Index

Man O'War *horse* 123
Manager *dog* 89, 90
Mancha *horse* 117
Mandy *dog* 6
Mangle *dog* 89
Mango *cat* 24
Marcus *dog* 7, 94
Marengo *horse* 123
Margaret *dog* 58; *sheep* 103
Margate *cat* 123
Marie *cow* 99
Marigold *cat* 79; *cow* 98
Marine Parade *horse* 83
Mariner *calf* 101
Marion *sheep* 103
Marjorie *sheep* 103
Marksman *dog* 89, 90
Marnie *dog* 52
Marocco *horse* 123
Marquin *dog* 69
Marshall *dog* 69
Marta *dog* 39
Martha *lamb* 103
martin 13, 19
Martin *monkey* 14, 15
Martine *dog* 53
Marty *dog* 48
Mary *cow* 99; *lamb* 103
Mascot *dog* 69
Masey *horse* 39
Masher *cat* 79; *dog* 94
Master Bristles *dog* 73
Master Magrath *dog* 123
Master Scorchin *horse* 84
Matchbox *dog* 89
Matchem *horse* 123
Mater *cat* 79
Mathe *dog* 123
Mathias *canary* 106
Matilda *cat* 45; *cow* 99; *praying mantis* 46
Maulkin *dog* 92
Maverick *cow* 103
Max *cat* 7; *dog* 6, 7, 8, 94
Maxi *dog* 64; *mouse* 61
Maxine *cat* 79, 80
May *lamb* 103
May Day *pony* 39
Mazawattee *cat* 79
Me *gosling* 60
Me Tarzan *horse* 85
Meadow Lady *horse* 84
Measles *pony* 26
Medina Boy *horse* 84
Meeny *goldfish* 60
Meg *cow* 99; *dog* 7, 72
Megan *dog* 89; *sheep* 103
Meggy *dog* 116
Mehitabel *cat* 154
Mel *dog* 32
Melanie *dog* 59
Melba *dog* 89
Melbourne *kangaroo rat* 58

Melita *dog* 74
Mell *cat* 58
Melody *dog* 69
Meo *cat* 79
Mercedes *dog* 62
Mercy *dog* 89
Merrishaw Rosamund *dog* 70
Merry *dog* 41
Merry Boy *dog* 92; *horse* 84
Merry King *dog* 57
Merry Prince *dog* 74
Merrylegs *dog* 154
Merryman *dog* 89
Messmate *dog* 89
Micetto *cat* 20
Michael *guinea pig* 53
Michelle *dog* 8
Micia *cat* 20
Micino *cat* 20
Micio *cat* 20
Mickey *cat* 7; *dog* 6; *mouse* 8, 9
Mickey Mouse *mouse* 137, 154
Microbe *dog* 74
Midi *mouse* 61
Midnight *cow* 99; *dog* 92; *horse* 123
Might *horse* 93
Mighty *horse* 93
Mighty Atom *cat* 79
Mighty Mouse *mouse* 137
Mignonne *cat* 20
Mijbil *otter* 124
Mike *dog* 116
Mild *cat* 59
Mildew *dog* 32, 91
Mile-a-Minute *horse* 82
Military Medal *horse* 83
Mill Reef *horse* 124
Millie *cow* 99; *dog* 124
Milligan *cat* 59
Millman *cow* 100
Milou *dog* 154
Mimi *cat* 20
Mimosa *cat* 54
Mincora *horse* 88
Minet *cat* 20
Minette *cat* 20
Ming *dog* 29
Mini *dog* 62; *mouse* 61
Minigold *horse* 83
Minna Minna Mowbray *cat* 124
Minnamour *horse* 85
Minnie *mouse* 8, 9
Minnie Mouse *mouse* 154
Mino *dog* 63
Minon *cat* 20, 154
Minou *cat* 20
Minstrel *dog* 26
Mint *cow* 99
Minus *guinea pig* 38

Miny *goldfish* 60
Misery *dog* 91
Miss Bristles *dog* 73
Miss by Miles *horse* 84
Miss Matty *cat* 24
Miss Mostly Brown *pony* 101
Miss Piggy *pig* 138
Miss Staek *calf* 103
Miss Whitey *cat* 79
Miss Wiffins *dog* 74
Missy *cat* 7; *dog* 6
Mr. Craven *sheep* 101
Mr. Fing *cat* 124
Mr. Mistoffelees *cat* 136
Mr. Pecky *cock* 101
Mr. Pym *cat* 38
Mr. Rusty *tortoise* 35
Mr. Toad *toad* 136
Mistigris *cat* 20, 154
Misty *cat* 7, 27; *pony* 8
Misty Mountain *horse* 24
Mittens *cat* 21
Mitzi *cat* 21; *dog* 8
Mo *goldfish* 60
Moby-Dick *whale* 155
Model *cow* 98
Model Princess *horse* 82
Modern *cat* 58
Modestine *donkey* 124
Moggie *cat* 16
Mohawk *dog* 89
Moira *cow* 99
Moko *cat* 79
Mole *mole* 136
Molly *cow* 99; *calf* 103
Mona *cat* 79
Monadelphia *dog* 92
Moneke *monkey* 15
*monkey* 15
Monkey Brand *dog* 74
Monocle *cat* 27
Monsieur Toutou *dog* 112
Monte Carlo *dog* 40
Montmorency *dog* 155
Monty *dog* 40, 51
Moonstone *pony* 24
Mop *dog* 68
Mopsey *dog* 131
Mopsy *rabbit* 61
Morecombe *lamb* 103
Morris *dog* 62
Morzillo *horse* 124
Moscow *cat* 79
Most Appealing *horse* 83
Mostly *pony* 101
Motley *dog* 92
Mouche *cat* 106
Moulin Rouge *dog* 68
Moumoutte *cat* 20
Mountaincrest Sparkler *dog* 73
Mountaincrest Welsh Spark *dog* 73

# Index 225

Mountaincrest Wonderboy *dog* 73
Mountbatten *cat* 55
Mouse *cat* 34
Mousie *mouse* 30
Mrs. Dick *cat* 31
Mrs. Nicola Hobbs *cat* 52
Mrs. X *horse* 102
Muckheap *cow* 103
Muffin *cat* 7; *dog* 6
Muffin the Mule *mule* 137
Mulberry *cow* 99
Mule Ears *dog* 125
Mungojerrie *cat* 136
Munkustrap *cat* 136
Murder *dog* 91
Muriel *cat* 54
Murphy *cat* 31
Murr *cat* 155
Murray the Outlaw of Fala Hill *dog* 115
Mushroom *cow* 98
Mushy *cat* 24
Music *dog* 74
Music Major *horse* 83
Musick *dog* 92
Musketeer *horse* 93
Mustang *dog* 69
Mustard *dog* 90; *lamb* 103
Mustavim *horse* 42
My Man O'War *horse* 123
Myrtle *pig* 100
Myth *dog* 63

Nadia *cow* 98
Nana *cow* 98; *dog* 155
Nancy *bullock* 101; *cow* 98, 99
Nanki Poo *dog* 40
Nannette *cow* 98
Nanny *goat* 13, 18
Napoleon *cat* 62; *hamster* 60; *pig* 155
Nasrullah *horse* 86
Natasha *cat* 50
Native Bride *horse* 85
Nautic *horse* 93
Nearco *horse* 86
Nectar *dog* 69
Neddy *donkey* 13, 18
Neeb *cat* 54
Negus *horse* 93
Nellie *cat* 27; *cow* 98; *horse* 93
Nelly *cow* 99
Nelson *cat* 27; *dog* 53; *duck* 103; *horse* 93, 125
Nemonie *dog* 69
Neptune *cat* 40
Nerissa *cat* 51; *dog* 69
Nettle *cow* 99; *dog* 70

New Forest *cow* 101
Newburgh *horse* 93, 94
Newes *polecat* 135
Newman *horse* 93, 94
Newsboy *horse* 93
Nibbles *hamster* 8
Nichola *horse* 93
Nicholas *cat* 52; *horse* 93
Nicholls *cat* 43
Nico *cat* 58
Nicol *rabbit* 53
Nicola *cat* 52; *cow* 98
Nijinsky *horse* 86, 125
Nijmegen *horse* 93
Nikki *cat* 48
Nikolai *cat* 50
Nim Chimpsky *chimpanzee* 125
Nimbus *horse* 93
Nimmo *sheep* 100
Nimrod *horse* 93
Nina *cow* 98; *dog* 69; *horse* 93; *lamb* 101
Ninja *cat* 60
Nino *dog* 63
Nipper *dog* 125
No Joke *dog* 73
Noble *horse* 93; *lion* 13, 14
Noddy-Tug *cat* 35
Nodnol *dog* 73
Noel *dog* 64
Noisy *dog* 74
Nona *cow* 98
Nonios *horse* 134
Nookie Bear *bear* 138
Noreen *cow* 98
Norfolk *horse* 93
Normandy *horse* 93
Norseman *horse* 93
North Star *horse* 93
Northern Baby *horse* 86
Northern Dancer *horse* 86, 125
Northern Taste *horse* 86
Northernette *horse* 86
Nosey Lad *pony* 34
Nubian *horse* 94
Nudder *dog* 33
Nuffie *cow* 97
Nugget *dog* 69
Numeral *horse* 94
Nutcracker *horse* 94
Nuts *lamb* 103
Nutty *dog* 41
Nylon *lamb* 100
Nyx *dog* 5

Obee *dog* 41
Oberon *dog* 69
Ocky *cat* 44
Octavius *cat* 44

Octo *dog* 40
Octopus *cat* 27
Oddsod of Doo Scallywag *dog* 73
Off Scent *horse* 85
Old Fred *pig* 101
Old Glory *rabbit* 25
Old Gold *dog* 69
Old Rowley *horse* 125
Old Yeller *dog* 155
Olga *cat* 48, 79
Oliver *cat* 34; *dog* 6, 69; *goose* 101
Oliver Rupert *cat* 64
Oliver Rupert Alexander Basil Randolph Quentin Bartholomew *cat* 64
Olivia *dog* 69
Omar *cat* 25
Omni *cat* 43
One I Lifted *cow* 102
Only a Monkey *horse* 83
Ophelia *dog* 69
Orbiter *dog* 69
Orchid Bud *dog* 69
Oremus *cat* 43
Oriel *dog* 69
Origen *dog* 69
Orlando *cat* 155
Orlon *horse* 85
Orville *animal* 138
Oscar *dog* 7
Oshkosh *dog* 125
Osprey *dog* 69
Oswald *dog* 69
Othello *lamb* 101
Otto *dog* 8
Ouch *goby* 60
Our Henry *horse* 84
Owd Bob *dog* 155
Owl *owl* 136
Ozzie *cat* 26

PJ *cat* 49
Paddington *bear* 156
Paddlepaws *cat* 27
Paddy *cat* 31, 39; *dog* 94
Pads *cat* 27
Pageant *dog* 89
Pagliacci *dog* 60
Palace Hope *horse* 82
Paladin *dog* 69
Pall Mall *horse* 86
Pam *lamb* 100
Pamela *dog* 89, 90
Pan *dog* 5, 156
Pancake *dog* 89
Pancho *dog* 98
Panda *cat* 25; *cow* 97
Panic *lovebird* 60
Pansy *cat* 79, 155; *cow* 99; *dog* 89

# Index

Pansy-Poo *cat* 33
Panther *dog* 89
Panzer *dog* 94
Papageno *dog* 60
Papillon *horse* 42
Paranoid Pam *cow* 99
Paris *cat* 55
Parity *dog* 89
Parker *dog* 45
Parsley *cat* 24
Parson *dog* 89, 90
Partlet *hen* 14, 15
Party Girl *horse* 84
Pask *dog* 40
Passion *dog* 89
Pastry *dog* 89
Patch *cow* 98; *guinea pig* 8
Patches *cat* 7
Patchwork *dog* 89
Patience *pony* 32
Patrick *cat* 54; *dog* 89
Patsy *cow* 99
Pave the Way *horse* 82
Pavlova *horse* 31
Peach *cow* 98
Peachy *cat* 62
Pearl *cow* 99
Pearl Button *dog* 57
Pecke in the Crowne *cat* 135
Pecky *cat* 32
Pecosa *dog* 112
Pedro *dog* 98
Peedicat *cat* 33
Peeka *cat* 58
Pegasus *horse* 156
Peggy *cow* 99; *dog* 116
Peggy Primrose *cat* 79
Pell *cat* 58
Peloton *dog* 107
Penates *cat* 58
Pini *dog* 29
Penguin *cat* 25
Penny *cow* 99; *dog* 6, 7, 33; *goldfish* 42; *lamb* 103
Pennycuik *horse* 83
Pensodoro *horse* 83
Pepe le Pew *skunk* 138
Pepi *dog* 8
Pepper *cat* 60, 79; *dog* 6; *lamb* 103
Percy *bull* 98; *cat* 35; *sheep* 100
Percy Bloggs *cat* 30
Perdita *dog* 69, 156
Peregrin *dog* 69
Perfect Marriage *horse* 85
Periander *cat* 112
Perky *cat* 30
Perruque *cat* 127
Perry *cat* 112
Persephone Smith *cat* 30
Persian War *horse* 126
Persimmon *cat* 79; *horse* 126
Pete the Pup *dog* 137

Peter *budgerigar* 8; *dog* 47, 94; *general* 2; *rabbit* 8, 9
Peter Pan *dog* 125
Peterman *horse* 87
Petite Etoile *horse* 126
Petra *dog* 126
Pewter *cow* 98
Phaeton *horse* 134
Phallus *horse* 126
Phantom *dog* 89
Phar Lap *horse* 126
Pharos *dog* 63
Pharosa *horse* 61
Philip *sparrow* 17
Philomel *dog* 69
Phlegon *horse* 134
Phoenix *dog* 63
Phread *parrot* 98
Phrenicos *horse* 126
Phyl-a-Pail *cow* 101
Phyllis *cow* 99
Physic *dog* 62
Pickles *cat* 57; *dog* 29, 63
Picnic *dog* 89; *pony* 59
Piddle *cat* 33
Pie *dog* 41, 58
Pierre *cat* 50; *dog* 6, 72
Pierrot *cat* 115
Pig *dog* 28
Piggie *pig* 100
Piglet *pig* 136
Pigolette *pig* 102
Pikeman *horse* 93
Pilgrim *dog* 89
Pillager *dog* 89
Pills *dog* 62
Pincer Teef *dog* 64
Pinch *guinea pig* 34
Pincher *dog* 5
Pini *dog* 29
Pink Rose *horse* 83
Pinky *cat* 27; *chaffinch* 35
Pinky and Perky *pigs* 138
Pinto *dog* 89
Pip *dog* 61, 129
Pippin *dog* 47
Pipps *cat* 25
Pittard *pig* 102
Pixie *cat* 79; *cow* 98
Planet *dog* 89
Planter *dog* 89, 90
Plato *dog* 92
Playboy *dog* 89
Playto *dog* 92
Pleader *dog* 89
Plonk *cat* 34
Ploughman *dog* 89, 90
Ploumeur Bodu *cat* 64
Plover *dog* 89
Plucky Punter *horse* 83
Plug *dog* 26
Pluto *cat* 111; *dog* 112, 138, 156

Poacher *dog* 89
Pobble *cat* 27
Pocket Picker *horse* 83
Pogo *opossum* 138
Pointer *dog* 163
Polestar *dog* 89
Polish *dog* 68
Pollux *cat* 59
Polly *calf* 103; *cat* 59; *cow* 99; *dog* 48, 68; *guinea pig* 25; *parrot* 13, 17, 18
Polly Wolly Wobbles *dog* 74
Polychrome *guinea pig* 25
Polynesia *parrot* 145
Pomegranate *cow* 98
Pomp *horse* 93
Pongo *dog* 156
Ponto *dog* 156
Poo *dog* 98
Poo-Chee *dog* 30
Pooh *swan* 165
Pooky *cat* 62
Pop *dog* 74
Popeye *dog* 27
Poppet *cat* 51; *horse* 100
Poppy *cat* 51; *cow* 99, 102; *general* 2; *dog* 30
Porchy *cat* 43
Porge *cat* 46
Porkie *pig* 100
Porky *dog* 28
Porky Pig *pig* 137
Portia *dog* 69
Portrait *dog* 89
Positive *dog* 89
Postman *dog* 89, 90
Posy *dog* 89; *guinea pig* 34
Potomac Volcano Silver Blackbird Blacky Sixpence *guinea pig* 65
Poultice *dog* 62
Powder Puff *cat* 79
Practical Pig, Fifer Pig, and Fiddler Pig *pigs* 157
Practice *dog* 89
Prejudice *horse* 93
Premium Bond *pig* 103
President *dog* 89, 90, 116
Priceless *dog* 89
Priddy Nice *horse* 83
Pride *horse* 83
Primrose *cow* 99; *dog* 89; *pig* 100
Prince *dog* 5, 6, 7, 10, 94; *horse* 100
Prince Charlie *cat* 79
Prince of Darkness *dog* 74
Princely Mount *horse* 82
Princess *cow* 98; *dog* 6
Pringy *dog* 57
Proctor *dog* 89, 90
Prodigal *dog* 89
Proggins *dog* 73

# Index

Promise *dog* 89; *pony* 41
Propel *horse* 82
Prospect *dog* 89
Prospero *dog* 79
Prowse *cat* 59
Prudence *cat* 45; *dog* 89, 125
Prudence Prim *dog* 125
Prudy *dog* 125
Prunella *dog* 47
Psalm *cow* 103
Psalmist *dog* 89
Psmith *dog* 50
Ptoophagos *dog* 133
Pudding *cat* 29, 64; *dog* 29, 58, 73
Puddles *dog* 33
Puff *cat* 29; *dog* 34
Puff Puff *cat* 42
Puffy *cat* 34
Pull-a-Pint *cow* 101
Punch *horse* 100; *lamb* 103; *pig* 98, 100
Punkin *cat* 7
Puppy *dog* 30
Pushinka *dog* 126
Pushkin *cat* 30
Puspots *cow* 102
Puss *cat* 19, 20, 27
Pussy *cat* 19, 20, 30
Pussy Cat *horse* 32, 38
Pussy-Jova *cat* 49
Pussy One Spot *cat* 60
Pussy Two Spot *cat* 60
Pussy Willow *cat* 25
Pyewacket *cat* 157
Pyewackett *cat* 135
Pynewacket *cat* 157
Pyrame *cat* 127
Pyramid *dog* 61
Pyrocis *horse* 134

Quaint Lady *dog* 69
Quaker *horse* 93
Quality Girl *dog* 69
Quartermaster *dog* 69
Queen *pig* 100
Queen Elizabeth *cat* 79
Queenie *dog* 6, 127
Queen's Treasure *horse* 82
Queer Street *dog* 74
Quelia *dog* 69
Quest *horse* 28
Queue Noire *cat* 26
Quick Thinking *horse* 83
Quicksilver *dog* 69
Quickstep *dog* 69
Quill *dog* 69
Quin *dog* 69
Quince *dog* 69
Quinote *dog* 69
Quip *cat* 27

Quite a Fool *dog* 73
Quite Sweet *horse* 83

Rab *dog* 127
Rabbit *rabbit* 136
Racan *cat* 127
Radegund *cow* 98
Rae *cow* 98
Rags and Tatters *dog* 74
Ragusa *cow* 98
Rain *pony* 40
Rain Lover *horse* 127
Rainbow *cow* 98; *horse* 26
Raine *cow* 98
Rajah *cat* 79; *dog* 94
Rakhsh *horse* 159
Raksha *wolf* 158
Raleigh's Revenge *dog* 38
Rally *dog* 37
Ram Jam *dog* 74
Rambo *rat* 8, 9
Ramsbottom *gosling* 60
Randy *dog* 73
Rani *cat* 28
Ranji *cat* 79
Rantan *dog* 69
Rapid River *horse* 82
Rapunzel *cow* 98
Raquela *cow* 98
Rat *cow* 102; *rat* 136
Ratty *rat* 8, 9, 30
Rawshark *dog* 25
Ready and Willing *horse* 85
Reagan *horse* 43
Real Jam *dog* 73
Realistic *horse* 83
Rebel *dog* 94
Rebella *cow* 98
Reckless *horse* 127
Red Alert *horse* 83
Red Cherry *dog* 69
Red China *horse* 83
Red Handed *horse* 87
Red Rum *horse* 127
Red Track *horse* 83
Redhead *cow* 98
Redwing *dog* 69
Regan *cow* 98; *dog* 69
Regina *cow* 98
Regula Baddun *horse* 158
Reine Beau *horse* 82
Relco *dog* 69
Reluctant Maid *horse* 84
Rembola *cow* 98
Rembrandt *cat* 38
Remus *cat* 43; *dog* 5, 57
Rene *cat* 55
Renuncula *cow* 98
Reuben *cat* 47
Rev. Dismal Doom *dog* 74
Rex *dog* 6, 94, 158; *horse* 127

Reynard *fox* 13, 14
Reynoldstown *horse* 128
Rezen *cow* 98
Rhapsody *dog* 69
Rheba *cow* 98
Rhoda *cow* 98
Ribo *dog* 51
Rica *cow* 98
Richelieu *cat* 79, 80
Richenda *cow* 98
Rickie *dog* 51
Ricky *dog* 94
Rienzi *horse* 128
Rigsby *dog* 48
Rik *cat* 51
Rikki-Tikki-Tavi *mongoose* 158
Rin Tin Tin *dog* 95, 128
Ringlet *cow* 98
Rinty *dog* 94, 95, 128
Rita *dog* 7
Roan *horse* 23
Roan Ranger *horse* 85
Roany *horse* 23
Roaring Wind *horse* 82
Rob Roy *dog* 125
Robbie *dog* 29, 52
Robert *horse* 93
Roberta *cow* 98
*robin* 13, 18
Robin *dog* 69
Robinson Crusoe *lamb* 103
Robyn *dog* 43
Rock *pony* 59
Rockefella *dog* 57
Rocky *cat* 7; *dog* 6, 10, 94; *horse* 29; *tortoise* 29
Roderica *cow* 98
Rodrigo de Triano *horse* 128
Roger *dog* 5; *rabbit* 8, 9
Roger Rabbit *rabbit* 137
Roiall Fluffball *cat* 79
Roland *rat* 8, 9
Roland Rat *rat* 9, 137
Roll *pony* 59
Rollo *cow* 98
Roma *cow* 98
Romance *cow* 98
Romany *cow* 98; *horse* 31
Romeo *dog* 69
Rommel *dog* 94
Romulus *dog* 57
Rondo's Boy *horse* 84
Ronnie *cat* 54
Roo *kangaroo* 136
Roobarb and Custard *dog and cat* 138
Roosevelt *dog* 158
Rosabelle *horse* 128
Rosalind *cow* 98; *dog* 69
Rose *cat* 59; *cow* 98, 99
Rose Pink *horse* 83

# Index

Rosebud *dog* 69
Rosetta *guinea pig* 28
Rosie *cow* 98, 99; *dog* 47; *guinea pig* 8, 9; *rat* 9
Rosinante *horse* 135, 159
Ross *dog* 7
Roswal *dog* 159
Rosy *cow* 99
Rover *cat* 51; *dog* 21, 51, 132
Rowena *dog* 69
Rowf *dog* 159
Rowlock *dog* 7
Roxanne *cow* 98; *pony* 41
Royal *horse* 93
Royal Dandy *horse* 83
Royale *cow* 98
Royda *cow* 98
Rozavel Golden Eagle *dog* 62
Ruby *cow* 99; *dog* 91
Ruby Rough *dog* 125
Rudi *cat* 32
Rue *cow* 98
Ruff and Reddy *dog and cat* 138
Ruffie *cat* 79
Ruksh *horse* 159
Rumpel *cat* 128
Rumpelteazer *cat* 136
Running Fire *horse* 82
Runsie *cat* 48
Rupert *dog* 68; *pony* 50
Rupert Bear *bear* 138
Russ *general* 23
Russet *dog* 69
Rustic Lad *horse* 84
Rusty *cat* 25, 76; *cow* 102; *dog* 6, 94; *general* 23
Ruth *cow* 99
Rutterkin *cat* 136
Rybolov *cat* 128

Sable *cow* 98
Sabre *dog* 94
Sabrina *cow* 99
Sacha *cat* 64; *dog* 49
Sacke and Sugar *rabbit* 135
Safety Blade *horse* 94
Saffron *cow* 98; *dog* 69
Sailor *dog* 89, 90; *horse* 93
Saint *dog* 39
Saintly Miss *horse* 84
Salar *salmon* 159
Salesman *dog* 94
Salisyn's Macduff *dog* 72
Sally *cow* 99; *dog* 7, 8, 52
Salote *dog* 39
Salt *cat* 60; *lamb* 103
Sam *cat* 7, 8, 9; *dog* 6, 7, 8, 94; *goldfish* 8, 9; *horse* 100; *spider* 47; *stick insect* 8, 9

Samantha *cat* 7, 47; *cow* 98
Sammy *cat* 7
Sampson *dog* 89
Samson *pony* 29
San Francisco *dog* 159
Sanctity *cow* 102
Sandra *cat* 42
Sandy *cat* 34; *dog* 1, 6, 72, 94; *general* 23; *gerbil* 9; *guinea pig* 8; *hamster* 24
Sangster *dog* 67
Santa *cat* 27
Santana *dog* 67
Sapphire *cat* 62; *dog* 69
Sara *lamb* 100
Saracen *dog* 90
Sarah *cow* 99
Sardar *horse* 128
Sasha *dog* 8
Satan *cat* 115; *dog* 27; *pony* 32
Satchmo *dog* 67
Satin *lamb* 100
Sausage *dog* 8
Savage Sam *dog* 155
Savoy *horse* 129
Saxon *dog* 69
Scally *dog* 73
Scallywag *dog* 73
Scarlet *dog* 35
Sceptre *dog* 69
Schubert *dog* 68
Schumann *dog* 34
Scilly *cow* 97
Scooby-Doo *dog* 138, 159
Scoot *cat* 33
Scout *pony* 41
Scraps *dog* 159
Screwy Bluey *horse* 28
Scrubber *dog* 68
Scylla *donkey* 59
Sea Prince *horse* 83
Searcher *dog* 132
Sebastian *dog* 68; *horse* 32
Sec *cat* 61
Secombe *dog* 67
Secret *dog* 74; *pony* 39
Secret Dream *horse* 85
Seductive *horse* 85
Sefton *cat* 40
Selaw *dog* 73
Selene *horse* 119
Selima *cat* 129
Selina *dog* 68
Selma *cat* 129
Semprini *dog* 67
Sentinel *dog* 69
Septimus *dog* 68
Séraphita *cat* 115
Serenade *dog* 69
Sergeant Rose *horse* 83
Sesquipedalius *cat* 27
Shackle *dog* 68

Shadow *dog* 63, 141; *horse* 24
Shadrach *dog* 68
Shamus *dog* 68
Shandy *horse* 24; *pony* 32
Shane *dog* 47, 94
Shannon *dog* 126
Sharky *whale* 10
Shasta *liger* 129
Sheba *cock* 57; *dog* 6, 7, 8, 51, 68
Sheldrake *dog* 89
Shelley *cat* 30
Shep *dog* 94, 95
Shere Khan *tiger* 160
Shergar *horse* 129
Sherry *dog* 41, 57, 63; *general* 23
Shiner *dog* 27, 68
Shirt Button *dog* 57
Shock *dog* 22
Short *horse* 100
Shorty *dog* 8
Shot in the Dark *horse* 83
Shotover *dog* 73
Shreddy *guinea pig* 38
Shrover *cat* 79
Shuffles *dog* 74
Si *cat* 79
Siam *cat* 79
Sid *cat* 55
Sideney *tortoise* 45
Silas *dog* 50
Silesia *dog* 50
Silky *lamb* 100
Silver *general* 23; *horse* 160
Silver Blaze *horse* 160
Silver Soot *cat* 79
Silver Strand *horse* 83
Silver Tiger *horse* 83
Silverdeen Emblem O'Neill *dog* 108
Simba *dog* 8, 94, 95, 129
Simon *dog* 51
Simpson *dog* 30
Sin *cat* 61
Singleton *sheep* 100
Singwell *dog* 92
Sinthia *cat* 33
Sir Harry *cat* 52
Sir Loyne *dog* 74
Sirius *dog* 29, 160
Skinfaxi *horse* 134
Skinhead *cow* 102
Skulker *dog* 160
Slap Bang *dog* 73
Slasher *cow* 102
Sleipnir *horse* 134
Slippers *dog* 144
Slurp *gourami* 60
Smacker *gourami* 60
Smackers *horse* 83, 86
Smart Sam *horse* 83

## Index

Smiler *horse* 100
Smith *cat* 50
Smokey *cat* 7, 8; *dog* 6; *rabbit* 8
Smokey's Girl *horse* 84
Smoky *cow* 97; *dog* 63
Smudge *general* 23
Snagglepuss *lion* 138
Snaggles *cat* 47
Snapdragon *cat* 79
Snarf *dog* 160
Snarleyow *horse* 161
Snarleyyow *dog* 161
Snatch *lamb* 100
Snitch *lamb* 100
Snitter *dog* 161
Snittle Timbery *dog* 130
Snoops *dog* 28
Snoopy *dog* 6, 7, 28, 72, 138, 161
Snow *cat* 51
Snowball *dog* 91; *pig* 92, 161
Snowdrift of Beetop *dog* 70
Snowdrop *cow* 99
Snowy *budgerigar* 8; *cat* 7; *dog* 154; *general* 23; *gerbil* 8; *guinea pig* 8; *hamster* 8; *pony* 9; *rabbit* 8
Snudge *sheep* 103
Snuff *dog* 24
So Valiant *horse* 83
Soames *cat* 50
Soap *dog* 68
Socks *general* 26
Soda *cat* 57; *dog* 58; *lamb* 103
Soldier *dog* 94
Soldier Boy *dog* 68
Solicitor *dog* 68
Solicitude *dog* 68
Soliloquy *dog* 68
Solitaire *dog* 68
Solitude *dog* 68
Solo *dog* 34, 67, 68; *sheep* 100
Solstar *dog* 68
Sonata *dog* 69
Sonya *cat* 50
Sooty *bear* 138; *cat* 7, 8 10; *general* 23; *gerbil* 8, 10; *guinea pig* 8, 10; *rabbit* 8, 10; *rat* 9, 10
Sophie *dog* 6, 7
Sorrel *horse* 23
Sos *cat* 44
Soumise *cat* 127
Soundwell *dog* 92
Sovereign *horse* 93
Sox *general* 63; *horse* 26
Space Shot *horse* 82
Spanish Gold *horse* 83
Sparkie *dog* 63, 73
Sparky *budgerigar* 40; *dog* 6, 40
Spartan *dog* 89
Specs *lovebird* 27
Speed Cop *horse* 82
Speedy *mouse* 8, 10; *tortoise* 8
Spice *pony* 38
Spike *cat* 59; *dog* 161
Spike and Tyke *dogs* 138
Spinnaker *dog* 68
Spit *dog* 68, 138
Spitfire *cat* 32; *dog* 92
Splash *horse* 26
Splinter *rat* 8, 10
Splodge *general* 23
Spock *cat* 28
Sponge *dog* 68
Spoof *dog* 74
Spot *cow* 97; *dog* 21, 70, 72
Spotlight *dog* 68
Spots *pig* 100
Spotty *general* 23
Spring Azure *horse* 83
Spring Secret *horse* 85
Spunky *dog* 119
Sputnik *dog* 92
Sqeaky *cat* 35
Squeak *dog* 61; *gerbil* 8, 9; *guinea pig* 8, 9; *lamb* 103; *mouse* 8, 9; *rat* 8, 9
Squeegee *dog* 68
Squeeker *pig* 102
Squirrel *cat* 27
Sredni Vashtar *ferret* 161
Stamp *gerbil* 60
Stanley *cockatoo* 33; *seagull* 52
Star *lamb* 103; *pony* 8
Starmeed Saturn's Siren *dog* 68
Starmeed Spectrum *dog* 68
Starshine *dog* 69
Starsky *rabbit* 49
Stenterello *dog* 161
Sterling Lad *horse* 83
Stick *stick insect* 8
Sticky *stick insect* 8
Stiffie *lamb* 100
Sting *goby* 60
Stirling *dog* 67
Stoneo Brokeo *dog* 74
Storm *horse* 39
Stormer *horse* 82
Straby *cat* 38
Straight as a Die *horse* 83
Stranger in Paradise *sheep* 102
Stravinsky *tortoise* 40
Strawberry *cow* 99; *horse* 23
Streaker *dog* 126
Streaky *general* 23
Strelka *dog* 107, 126
Strider *whale* 10
Strike *dog* 94
Stumah *dog* 161
Stumblebum *dog* 31
Stumps *cat* 79
Stumpy *cat* 27
Stunning *horse* 83
Stupendous Boy *horse* 84
Stutterford *cat* 52
Sue *cow* 99; *dinosaur* 129; *dog* 47
Sugar *dog* 63
Sugar Ray *horse* 31
Sukie *cat* 59
Suky *hen* 24
Summer Queen *horse* 40
Sun Queen *horse* 83
Sunday *pony* 39
Sunflower *dog* 69
Sunny Jim *dog* 37
Sunnyboy *horse* 84
Sunreef Harvest Glow *dog* 72
Sunshine *cow* 99; *pony* 34
Surprise *pig* 100; *pony* 39
Surrey *dog* 28
Surtees *dog* 67
Susa *cat* 79
Susan *cat* 79; *dog* 63
Susan Sow *pig* 102
Susie *cow* 99; *dog* 6, 7
Sutton Bonington *cow* 102
Suzette *dog* 8
Suzuki *dog* 30
Svadilfari *horse* 134
Swan *cow* 98
Sweep *cat* 62; *gerbil* 9, 10
Sweep the Green *dog* 73
Sweeper *dog* 68
Sweet Lips *dog* 132
Sweet Slavery *horse* 85
Sweetheart *dog* 161
Swell Fellow *horse* 83
Swiftly *dog* 73
Swisher *horse* 31
Sydney *kangaroo rat* 58
Sykes *dog* 67
Sylvanecte *horse* 85
Sylvester *cat* 10, 138
Sylvio *dog* 161
Symphony *dog* 68, 69

T.C. *cat* 163
T. Wilson, Esq., M.P. *cat* 64
Tabby *cat* 7, 61
Tabbyshankers *cat* 50
Tabitha *cat* 7
Tachebrune *horse* 161
Tackle *dog* 68
Tad *cat* 27
Taffy *dog* 6
Tags *cat* 79, 80
Talbot Flicker *dog* 102

# Index

Tally *dog* 57
Tamarind *dog* 69
Tamarisk *pony* 24
Tammy *cat* 61; *pony* 24
Tamora *dog* 69
Tan *cat* 61
Tangle *cat* 53
Tanky *pony* 29
*tantony pig* 19
Tao *cat* 130
Tara *dog* 7, 8
Tarantella *horse* 31
Tarka *otter* 162
Tarot *cat* 41
Tartar *dog* 162
Tash *cat* 27
Taster *dog* 132
Tattler *dog* 92
Tawny *general* 23
Teaspoon *cat* 34
*teddy bear* 22
Teddy *cow* 102
Teddy Tail *mouse* 137
Tee-Vee *dog* 92
Teem *dog* 162
Tempered *horse* 83
Terrible Tim *dog* 125
Terry *tortoise* 8, 9
Tess *dog* 47
Tessa *cow* 99; *dog* 6, 7
Test *dog* 162
Thanksgiving *calf* 102
The Absent-Minded Beggar *cat* 79
The Admiral *horse* 83
The Babe *dog* 74
The Baldies *hens* 102
The Cadger *dog* 74
The Dawn Chorus *pigs* 102
The Duke *dog* 63
The General *cat* 26
The Ghost *dog* 74
The Hairy King *dog* 74, 75
The Microbe *dog* 74
The Mighty Atom *cat* 79
The One I Lifted *cow* 102
The Pop *dog* 74
The Proggins *dog* 73
Theo *cat* 43
Thisbé *cat* 127
Thor *dog* 39
Three Crowns *horse* 83
Three-to-Nothing Jack Dalton *goat* 107
Throttler *dog* 160
Thumper *rabbit* 8, 10
Thunder *dog* 92, 94; *general* 59; *horse* 59, 93
Ti Ping *dog* 30
Tibbles *cat* 14
Tibby *cat* 61
Tibert *cat* 14
Tickler *dog* 92

Tiddles *cat* 14, 130
Tiddly-Pom *cat* 62
Tiddly-Poos *cat* 33
Tiercelin *rook* 14, 15
Tiger *cat* 7, 8, 130; *dog* 6
Tiger Tim *tiger* 137
Tigger *cat* 7, 8, 58; *tiger* 136
Tiggy *cat* 25; *dog* 31
Tiggy-Wee *cat* 51
Tille *cat* 59
Tiller *dog* 68
Tillsome Puss *cat* 64
Tilly *cat* 47, 64
Tilly Williams *cat* 64
Tilly Willow *cat* 64
Tilly Willy *cat* 64
Tilly Wimpole *cat* 64
Tide *horse* 93
Tim *dog* 121
Timber Doodle *dog* 130
Time *horse* 83
Timkins *cat* 79
Timmy *cat* 1, 53, 62; *tortoise* 8, 9
Timolin *horse* 42
Timon *dog* 69
Timothy White *cat* 53
Tingha and Tucka *koalas* 138
Tinker *cat* 30, 47, 50; *dog* 63
Tinkerbell *cat* 30
Tinker's Trade of Harmaur *dog* 73
Tinki-Tonk *cat* 30
Tinkle *cat* 155
Tiny *cat* 28; *cow* 97; *dog* 63
Tiny Tim *dog* 125
Tip *dog* 26
Tippler *dog* 132
Tippy *dog* 6
Tipsy *hamster* 38
Tishy *cat* 38
Titania *dog* 69
Tito Gobbi *dog* 34
Tizzy *cat* 61
Tobby *cat* 61
Toby *cat* 7; *dog* 6, 7, 8, 162; *lamb* 100; *pig* 100; *tortoise* 8, 9
Toby-Chien *dog* 163
Toby Tup *sheep* 102
Tod *dog* 50
Todd *cat* 34
Toffee *dog* 8, 23
Token Girl *horse* 84
Tom *cat* 8, 9; *general* 60; *gerbil* 8, 9; *goldfish* 8, 9; *lamb* 103; *mouse* 8, 9; *rat* 9; *stick insect* 8, 9
Tom and Jerry *cat and mouse* 138, 163
*tom cat* 13, 18
Tom Jones *lamb* 102
Tom Quartz *cat* 130

Tomahawk *dog* 94
Tomboy *dog* 91
Tommy *tortoise* 8, 9
Toms *cat* 58
Tonic *dog* 62
Ton-Kee *cat* 30
Tony *dog* 5; *horse* 130
Top Cat *cat* 138, 163
Top Notch *whale* 10
Top Speed *horse* 82
Topaze *cat* 62
Tophat *cat* 27
Topic *pony* 59
Topper *dog* 73
Topso *cat* 79
Topsy *cat* 38, 51, 79; *cow* 97; *dog* 131; *donkey* 130; *sheep* 100
Tortie *tortoise* 47
Torty *cat* 30; *tortoise* 30
Tory *dog* 62
Tosca *dog* 48
Tosh *goldfish* 28
Tosqua *cat* 60
Toss *cat* 106
To-To *cat* 79
Toto *dog* 163
Touchstone *dog* 92
Tower Bridge *dog* 74
Towzer *dog* 21
Tracey *dog* 8
Transom *dog* 89, 90
Traveller *horse* 131
Tray *dog* 121, 163
Treacle *dog* 8
Trebizond *horse* 164
Trepp *dog* 131
Trident *horse* 93
Trigger *horse* 131
Trimmer *horse* 100
Trinidad *horse* 39
Tripod *cat* 27
Tristram Shandy *pony* 32
Trixie *dog* 6, 47, 72
Trojan *dog* 94
Trotty *dog* 57
True North *horse* 86
Truelips *dog* 92
Truelove *dog* 89
Tudor Shoon *horse* 83
Tufnell *cat* 55
Tulip *cat* 29; *dog* 127
Tunis *horse* 131
Tupelo *cat* 49
Turk *dog* 115
Turpin *horse* 100
Tusky *cat* 61
Tut *cow* 102
Tu-Tu *cat* 40
Tweed *lamb* 100
Tweedie *cat* 64
Tweetness *dog* 153
Tweety *budgie* 8, 10

## Index

Tweety Pie *bird* 10; *canary* 138
Twiddely-Poof *cat* 64
Twiddely-Toes *cat* 64
Twigger *cats* 58
Twiggy *cat* 62; *pony* 28; *stick insect* 8
Twin *cat* 79
Twinkle *cow* 98, 99; *lamb* 103
Twinkletoes *cat* 26
Twitch *cat* 62
Twitch-Three *cat* 62
Twitch-Three-B *cat* 62
Twitch-Two *cat* 62
Twizzle *cat* 31
Two for Joy *horse* 85
Twopence *lamb* 103
Tyras *dog* 131
Tyson *dog* 10

Ukey *cat* 43
Ulster *horse* 93
Unbiased *horse* 83
Uncle Cyril *horse* 84
Union *horse* 93
Up to Date *dog* 74

Vale *dog* 58
Valentine *cat* 48
Vandyke *cat* 28
Vanessa *cat* 49
Vatellor *horse* 87
Velvet *sheep* 100
Venture *horse* 100
Venus *cat* 131
Verdant Green *horse* 83
Vic *dog* 62, 70
Vicar of Leeds *dog* 74
Victor's Hussar *horse* 83
Victory *dog* 92
Vim *rabbit* 60
Vinegar Tom *dog* 135
Violet *cow* 99; *horse* 100
Viv *cow* 99
Vivace *dog* 68
Vodeira *horse* 88
Vonolel *horse* 131
Vulcan *dog* 131

Waif *dog* 4
Waifer *cat* 38
Wakewood *dog* 90
Walbrook *horse* 93
Wales *horse* 93
Wallflower *cat* 79

Wally Pug *cat* 79
Wanda *angel fish* 164
Warrior *dog* 90
Warspite *dog* 94
Washaway *horse* 42
Washington *horse* 93
Washoe *chimp* 132
Waterloo *dog* 92
Wattle *duck* 60
Watty *cow* 99
Wee Wee *dog* 74
Weed *pony* 28
Weedy *dog* 68
Weejie *dog* 121
Wellington *cat* 26; *dog* 92
Wendy *cow* 99; *horse* 53
Wessex *dog* 132
Wettex *dog* 68
What Care I *dog* 73
What's Wanted *dog* 73
Whisker *pony* 164
Whisky *cat* 7, 26, 28, 34, 43; *dog* 58, 63; *general* 23; *lamb* 103
Whissendine *horse* 93
White *cow* 98
White Fang *wolf-dog* 164
White Friar *cat* 79
White Hope *horse* 83
White Lady *cat* 76
White Prince *horse* 83
White Surrey *horse* 165
White Tip *dog* 126
White to Move *horse* 83
Whitefoot *horse* 100
Whitey *general* 23
Whitie *general* 23
Whity *general* 23
Whoomph *horse* 82
Wigger *cat* 58
Wiggins *pig* 100
Wikket *cat* 51
Wilberforce *cat* 119, 132
Wilbur *pig* 165
Wildnight *dog* 90
Wildrose *dog* 90
Wilfred *dog* 61
Will Go *horse* 86
William *dog* 6, 7
Willie My Son *horse* 84
Willis *cat* 61
Willow *dog* 90; *horse* 42
Willy-Boy *sheep* 100
Willy Wet *dog* 33
Wilson *cat* 64
Wimothy Tight *cat* 53
Wimpole *cat* 63
Windmill *pony* 38
Windsor Lad *horse* 87
Windy *dog* 30
Windy Dick *dog* 74
Winifred *cat* 45

Winkie *dog* 5; *hamster* 27
Winks *cat* 79
Winnie *bear* 165
Winnie-the-Pooh *bear* 165
Winning Hand *horse* 83
Winsome *horse* 83
Winston *horse* 93, 132
Winter Moon *pony* 39
Wise *lamb* 103
Wise Virgin *dog* 74
Wolf *dog* 94, 127
Wolfgang *dog* 74
Wonder *pig* 100
Woody *cow* 102
Woody Woodpecker *woodpecker* 137
Woofe *dog* 35
Woofer *dog* 62
Woo-Hoo *cat* 28
Woolly *lamb* 100
Wooloomooloo *cat* 79
Woomsey *rabbit* 60
Wooshams *cat* 31
Woosie *dog* 5
Wopsey *rabbit* 60
Wot Av I Mist *horse* 83
Wowski *dog* 91
Wren *dog* 41

Xanthus *horse* 134
Xerxes *horse* 93

Yackie *cockatoo* 35
Yam Seng *guinea pig* 38
Yarraman *horse* 30
Yatzi *dog* 40
Yawl *dog* 68
Yogi *dog* 50
Yogi Bear *bear* 138, 165
Yolanda *cat* 79
You *cat* 47
Young Cato *horse* 84
Young Robert *horse* 84
Yo-Yo *cat* 31
Yuki *dog* 132
Yukon *dog* 121
Yula *cat* 79, 80
Yum Yum *dog* 74
Yze *dog* 73

Zac *goose* 102
Zebedee *bee* 147; *sheep* 100
Zoe *dog* 6, 7
Zoroaster *cat* 106
Zulu *cow* 99
Zuzie *dog* 47